T0261308

# Science, Religion,
# and Mormon Cosmology

and God said:

$$p = mv$$

$$F = \frac{dp}{dt}$$

$$F = G\frac{m_1 m_2}{r^2}$$

$$v = H_0 d$$

$$\oint E dA = \frac{q}{\epsilon_0}$$

$$\oint B dA = 0$$

$$\oint E dl = -\frac{d\Phi_B}{dt}$$

$$\oint B dl = \mu_0 l + \epsilon_0 \mu_0 \frac{d\Phi_E}{dt}$$

and there was

LIGHT!

# Science, Religion, and Mormon Cosmology

Erich Robert Paul

UNIVERSITY OF ILLINOIS PRESS

*Urbana and Chicago*

© 1992 by the Board of Trustees of the University of Illinois
Manufactured in the United States of America
C  6  5  4  3  2

*This book is printed on acid-free paper.*

The author gratefully acknowledges permission to reproduce two illustrations from
Andreas Cellarius's *Harmonia Macracosmica* and of Galileo with the Muses, all published
in Hugh Kearney, Science and Change, 1500-1700. Copyright © 1971 by McGraw-Hill.

Library of Congress Cataloging-in-Publication Data

Paul, Erich Robert, 1943–
Science, religion, and Mormon cosmology / Erich Robert Paul
p.  cm.
ISBN 10: 0-252-01895-8 (alk. paper)/ISBN 13: 978-0-252-01895-4
1. Cosmology, Mormon. 2. Religion and science—History. 3. Mormon
church—doctrines. 4. Church of Jesus Christ of Latter-day Saints—Doctrines. I.
title.
BX8643.C68P38  1992
261.5'088283—dc20                                                    91-30916
                                                                          CIP

*To Kathleen*

# Contents

*Illustrations follow pages 54, 112, 134, and 158.*

# Preface

This study originated in two unrelated events. My developing interest in the history of scientific cosmology began in graduate school nearly two decades ago and has culminated in the forthcoming publication of my *Milky Way Galaxy and Statistical Cosmology, 1890–1924* (Cambridge University Press). Exploring the nature of scientific cosmology during the nineteenth and early twentieth centuries, this book provides the background for many of the scientific issues discussed here. As I worked on it, I realized that Mormonism could provide a unique case study in the relationship of science and religion. First, Mormon theology entails numerous philosophical and theological claims that can be construed as favoring a scientism. And second, Mormonism has wedded ideas on cosmology to its very theological fabric.

My purpose here is to explore historically the relationship of science and Mormonism. Although there are many, perhaps even equally legitimate, ways to focus this study, I have chosen the idea of *cosmology* as my primary theme because I genuinely believe that as an organizing principle, cosmology is the most basic conceptual means for exploring this complex relationship. My discussion is not primarily an excursion into Mormon theology. While I constantly refrain from explicitly examining this much larger issue, I argue that the idea of cosmology is one central organizing theme in Mormonism.

Because this study is equally the result of my thinking about Mormonism and my investigation of the relationship of science to religion, as well as the product of a particular research project, over many years I have benefited from searching and enlightening discussions with Mormon scholars Douglas D. Alder, Edward R. Hogan, Hollis R. Johnson,

Robert L. Miller, Sheldon T. Miller, and Jan Shipps and also with fellow historians of science Michael J. Crowe, Edward B. Davis, Steven J. Dick, and Jane Maienschein, who have deeply enriched my understanding of the development of science. With deepest gratitude I acknowledge my intellectual mentors Edward Grant, Richard S. Westfall, and the late Victor E. Thoren, who influenced me profoundly during my years as a graduate student at Indiana University.

Numerous professional colleagues have contributed to this book by generously giving advice and/or criticisms. Doug Alder, historian and president of Dixie College, provided invaluable comments throughout and caused me much pause. David H. Bailey of NASA-Ames Research Center read the entire book, providing particularly helpful suggestions for chapters 8 and 9, and constantly provoked me to deeper reflection. Michael Crowe of the University of Notre Dame, who was initially responsible for urging my interests in Mormon cosmology, provided indispensable suggestions on chapters 1 through 5 and allowed me access to his own research on the 'plurality of worlds'. Edward Davis of Messiah College provided helpful comments on chapters 1 through 3. I am indebted to Steve Dick, senior historian at the U.S. Naval Observatory, for his invaluable comments on chapter 9 and his discussion of the 'biophysical cosmology'. Edward Hogan of East Stroudsburg University read the entire book, providing rich and detailed commentary, particularly on chapter 6. Zoologist Duane Jeffery of Brigham Young University also read the entire book, contributed extremely helpful suggestions on evolutionary issues in chapters 8 and 9, and shared with me his rich experience in evolutionary studies. Astrophysicist Hollis Johnson of Indiana University, who initially stimulated my thinking about SETI and possible Mormon connections, and Von Del Chamberlain of the Hansen Planetarium provided insightful comments to chapter 9. Robert Miller at the Marriott Library of the University of Utah allowed me full access to his bibliography and notes on Mormonism and science, saving me additional months of searching primary and esoteric sources. Finally, I am grateful to Thomas G. Alexander, historian at Brigham Young University, and to Hollis Johnson, both of whom, in anonymously reviewing the entire manuscript for the University of Illinois Press, challenged a number of assertions, thus giving me additional pause for reconsideration.

Numerous friends and colleagues have offered continuous encouragement and stimulation for many years. Thanks are due especially to Delbert W. Ellsworth, Gary W. Hansen, F. Kent Nielsen, Keith E. Peterman, and the late Richard M. Sheeley, all of whom provided valuable suggestions about the Mormon connections to science. Finally, I thank friends Robert Dixon, Howard Kempton, Allen Parris, Bill Paul, and Tom Williams, who have listened to my thinking on Mormonism as it has evolved. Above all, I thank Kathleen, who has been my severest, yet most compassionate, critic in matters of tone and balance.

Research assistance has been provided over the years by the staffs of a number of libraries, including the Boyd L. Spahr Library of Dickinson College, the Harold B. Lee Library and Special Collections of Brigham Young University, the Library of Congress Manuscript Reading Room, Dr. Greg Thompson and his staff in Special Collections at the Marriott Library of the University of Utah, the Historical Department of the Church of Jesus Christ of Latter-day Saints, and the Ontario County Historical Society in Canandaigua, New York. I am grateful to the editors of *Dialogue: A Journal of Mormon Thought* for permission to reproduce material from my previous articles. Cynthia Mitchell and the editorial department at the University of Illinois Press in Urbana have given invaluable aid in preparation of the final manuscript.

Finally, I am grateful to Dickinson College for a Board of Advisors research grant and to the Philosophy Department at Utah State University, both of which provided additional financial means necessary to complete this book during a recent year-long sabbatical in the Wasatch Mountains. Kathleen and our children, Ann-Marie, Lisa, Juliet, Erica, and Christopher, supported me during many hours of solitary reflection, occasionally providing relief as we explored issues of cosmic significance at Park City, Snowbird, and Alta.

While I am indebted to the wisdom of all these friends and colleagues, responsibility for any errors of fact or judgment that occur is entirely mine.

# Science, Religion, and Mormon Cosmology

Since its founding in 1830, Mormonism has not only evolved a new canon that set the religious movement apart from mainline Protestantism but also fashioned an ethos that embraces a positive scientism in keeping with the roots of Greek rationalism. Simultaneously, drawing deeply from both prophetic and Hebraic sources in the Judeo-Christian tradition, Mormonism has constructed a new understanding of human experience. Although this exceptionalist view has not gone unchallenged, interpreting Mormonism as a new religious tradition casts light on the movement and the construction of its cosmology.[1]

## Mormonism and Science

While Mormon cosmology has been shaped by a broad range of theologically interrelated ideas, many of its elements have been cast in scientific terms.[2] Historically, these can be organized into four general categories: (1) American antiquities and new-world archaeology; (2) technology; (3) Darwinism, organic evolution, and geological studies of the age of the Earth; and (4) astronomy and cosmology.

From Mormonism's inception, believers have responded enthusiastically to archaeological notions. In the process Mormonism has developed a lively tradition in Book of Mormon archaeology, which continues unabated to this day. While in earlier times the penchant for a literal-historic approach to the Book of Mormon vastly dominated, more recently, as some Mormons have become sophisticated scientifically, they have been more cautious in ascribing specific themes and events found in the Book of Mormon to various archaeological finds and views.[3] Although these studies are central to the meaning of the Mormon canon, for the present study they do not directly impinge on the broadest issues of science and religion.

In the minds of most Americans science and technology have become equated with progress. Consequently, Mormons, like Americans generally, see the achievements of technology—such as rail travel, telegraphy, and the telephone in the nineteenth century and airplanes, nuclear power, and computers in the twentieth—as equivalent to modern science. This confluence of ideas tends to confuse the *inventions* of technology as applied science with the underlying scientific *ideas*. Because technology, particularly prior to the First War, was largely benign and for most people did not threaten values and beliefs, the popular image of science has remained quite positive.[4] In large measure the church's view of science reflects these nineteenth-century attitudes.

Although the Darwinian redefinition of humanity's place in nature has affected all of Western thinking in deep and lasting ways, Darwinism, rather than cosmology, had a far greater impact on Protestants than on Mormons.[5] Although mainstream Christianity eventually made peace with Darwinism, in general Christians initially responded defensively to the emergence of organic evolution, partly because cosmology was never an integral part of Protestant theology, as it was (and is) for Mormonism, but also because Protestants before Darwin interpreted humankind's origins exclusively from a literal reading of Genesis whereas Mormons strongly augmented their views from contemporary revelation and other prophetic sources. This is not to suggest that evolutionary thinking has not affected Mormons.[6] Indeed, it has! For example, the theory of organic evolution is widely taught by Mormon scientists at Mormon colleges and universities. With the exception of only a few individuals, however, evolutionary views have had virtually no direct, positive effect on the writings of Mormon religious authorities (see chapter 8). Conversely, since early in the century to the present, a small number of quite vocal Mormon authorities have actively sought to discredit evolutionary thinking. Consequently, its effect has simply been slower and less dramatic than on Mormonism's religious neighbors, most of whom have accommodated themselves to evolutionary ideas.

Prior to the emergence of an evolutionary cosmic vision in the twentieth century, the very concepts that defined humankind's place in nature—'creationism', 'catastrophism', 'the young Earth', 'essentialism', 'anthropocentrism'—were themselves couched in the nineteenth-century holistic terms of a nondevelopmental and mechanical worldview. To be sure, historically, evolutionary thinking provided a set of ideas that eventually challenged basic religious presuppositions and claims.[7] But whereas a discussion of the origins of humankind may be central to Genesis, in the Mormon tradition Genesis also conveys a mythopoetic understanding of the origins of humankind. In this sense Genesis is cosmology first.[8] Thus I will interpret evolution broadly, as dealing with *origins*, and place it in the subdiscipline of cosmology known as cosmogony.

Some will argue that Mormonism's encounter with evolution—and not cosmology—is the central issue in science and Mormonism. For the Mormon experience there are three reasons why this is not so. First, much of Mormon cosmology had already been reasonably well developed before Darwin's ideas, first published in his *Origin of Species* in 1859, had become known. Indeed, the full implications of Darwinian evolution were not apparent to the popular audience until the end of the century. Second, after Darwinian evolution entered the marketplace of ideas on the American frontier, Mormons responded in ways that were not unlike their theological neighbors; namely, they uncritically rejected the new developmental notions—at least until 1909 (see chapter 8). And third, unlike evolution, the science and metaphysics of cosmology have been wedded to the very fabric of Mormon theology. Both as worldview and as science, cosmology may be the one area of the natural sciences that has the most to say about Mormonism and its relationship to science.[9]

On astronomical and cosmological matters, Mormonism historically cultivated a particularly active and positive interest. Although cosmology can be studied as a technical subdiscipline of physics and astronomy, it can also be understood loosely as worldview, as something that is perhaps more value-laden than scientific. I want to use the term to connote an understanding of the structure of both a physical and a transcendent universe. In this sense, both science and religion have something to say about the shape and texture of that cosmology.

## Cosmology and Mormon Theology

The term 'cosmology', which can be used broadly to subsume the study of origins known as cosmogony, is culturally the study of both our private and shared worldviews and technically the study of the architecture of the universe. This study examines historically how Mormonism shaped its cosmic vision by using and developing cosmological ideas and what this says about both science and religion. The cosmologies of some of the most prominent Mormon thinkers, particularly those of Joseph Smith, Brigham Young, Orson Pratt, and

B. H. Roberts, will be treated in various detail insofar as they relate to questions directly bearing on science and its relationship to Mormonism specifically and to religion generally.

Because of the centrality of these four individuals to the development of Mormon thought, their cosmologies significantly affected the emergence and development of a Mormon cosmic vision. Here their private cosmologies will be used as vehicles to explore Mormonism's relationship to the science of the day. In the broadest sense cosmology deals fundamentally with the private mindset of an individual. As such, the cosmology of someone like Joseph Smith is very complicated, including not only his whole range of ideas but also their origin. Consequently, I will use the term 'cosmology' in a restricted sense to apply to issues centered on the relationship of science and Mormonism.[10] Because the cosmologies of these individuals were cast within the same general framework, they reflect similar theological and philosophical assumptions. Consequently, it is possible to speak of a universal Mormon cosmology.

This cosmology has shaped and affected Mormonism's understanding of science in contemporary times. This study clearly shows that the emergence of this cosmology compelled Mormons to develop a positive perception of science precisely because they understood science first from within a theological frame. For example, such Mormon doctrines as the plurality of gods, premortal spiritual nature of humankind, 'degrees of glory', 'eternal progression', and particularly 'worlds without number' compelled Mormons to adopt an idiosyncratic but sympathetic understanding of cosmology. As a consequence, these views reflect a cosmology which assumes a metaphysical pluralism, adopts an eternalism, and urges a developmental (process) theology. Although some have tried, no one has ever succeeded in developing these views into a formal, doctrinaire, or systematic theology, nor will it be our purpose to do so. Alternatively, coupled with prophetic elements, a somewhat flexible theology has protected Mormonism from most epistemological and metaphysical dilemmas inherent in theologies less secure and more rigid.

Much of the theology expressed in the Mormon canon, particularly as found in the ideas enumerated above, has an eschatological orientation that revolves around the penultimate conditions of postmortal life. For Mormons these ideas found their quintessential expression in cosmological themes that appear in the church's most sacred ceremonies, where the microcosm of the universe, expressed in vivid detail, subsumes the plurality-of-worlds idea. With this one doctrinal exception, in numerous speeches, essays, and books, official and lay Mormon authorities have fashioned these concerns into a coherent but loosely defined understanding or, in Mormon parlence, a "plan."

## Mormonism and Its View of Science

The process of constructing a scientific cosmos has raised some significant philosophical and theological issues within Mormonism. For example, a realist conception of science—the view that there is a direct correspondence between ultimate reality and the perception of that reality sanctioned by scripture—has led largely to an uncritical acceptance of science. Moreover, although Mormons believe that God is ultimately the author of nature, Mormonism developed a view of 'natural law' which sanctioned not only the idea that nature stands largely independent from that creation but an understanding of nature which must first recognize the lawfulness of nature itself. Neither of these led to the development of a natural theology (theology deriving its knowledge of God from the study of nature rather than revelation), however, as opposed to revealed religion.

To be sure, Mormonism reflected many of the same problems and conflicts in the larger American religious experience. But with its roots embedded in Greek and Hebrew traditions, Mormonism's blending of these intellectual and spiritual dimensions resulted in the emergence of unique tensions. For example, as science changed sufficiently, Mormonism might have been unable to adjust its theological base to accommodate science. This was not the case, however, because Mormonism possesses a new canon, dominated by an epistemology that allows for the expression of prophetic views.

Until roughly the 1930s Mormonism was a vilified American religion. In the face of legal, social, and intellectual opposition, Mormonism not only survived, it flourished.[11] Although my study will not focus on this issue, the mere success of Mormonism, despite the odds and a sometimes close connection to science, compels an exploration of the relationship between science and Mormonism within the broader dimensions of the American religious experience. In considering this, I reject the warfare thesis—that there is some kind of inherently antagonistic relationship between science and religion—as being conceptually misdirected and historically largely false.

Alternatively, I accept the view that one way to understand this relationship is to explore religion from within, so to speak, as it exploited ideas drawn from the larger scientific milieu. For example, as Mormonism matured, the scientific (and theological) idea of the plurality of worlds—the claim that life exists on other world systems—

became a basic premise of its theology. At the time when scientific cosmogony and biblical creationism were the sources of increasing polarization within many Christian sects, many within Mormonism (despite considerable reaction and much literal-historic leanings within the movement) have still been able to accommodate themselves to an old-earth/ old-universe view. Mormonism rejected traditional *ex nihilo* accounts of the creation, suggesting instead that the universe is composed only of preexisting material substance. These and related issues suggest that Mormonism's post-Enlightenment interpretation of humans and nature compelled it to understand the ground of science in a literal, though somewhat embryonic, way. Still, although Mormonism developed a positive conception of science, it managed to avoid the kind of natural theology that challenged many Protestant sects.

The Mormon movement has reflected some aspects of science in its worldview. This suggests a range of issues. Was science lifted in toto to support the nascent Mormon theology? If so, did Mormonism develop a natural theology to support its version of religious understanding? Or did Mormonism accept only limited aspects of science and, if so, which parts and why? Did Mormonism develop a cosmology of the universe based largely on a theology that was only loosely supported by science? What has been the attitude toward science by Mormon theologians and intellectuals?

During Mormonism's first century, roughly the century before the Second World War, Mormon attitudes were shaped profoundly by scholar-authorities Parley P. and Orson Pratt, B. H. Roberts, James E. Talmage, John A. Widtsoe, and Joseph F. Merrill. Their legacy of affirming the value of knowledge for its own sake as a reflection of a divine imperative remains a powerful force within Mormonism. With the passing of these scholar-authorities, however, the exponential growth of the church has pressed much of its creative energy into institutional issues. Although the influence of Mormonism's first century remains a powerful living legacy, church authorities no longer direct and shape beliefs in these intellectual ways. Although considerable innovative energy still exists within Mormonism, now it is focused within the church's body politic. Mormonism's once highly creative intellectual presence has been partially replaced in recent years by a powerful but residual antiscientism and anti-intellectualism among some Mormons.

Although still seeking knowledge as an expression of some divine imperative, late twentieth-century Mormons have also adopted normative cultural values. These attitudes are neither unique nor an aberration within the Mormon culture. Indeed, a 1974 article appearing in *Science*—published by the largest scientific society in America, the American Association for the Advancement of Science, and, along with the British journal *Nature*, certainly the most influential science magazine—reported that Mormonism had produced more scientists per capita than virtually all religious movements in twentieth-century America.[12] Although there are social, religious, and theological reasons for this mostly supportive relationship, the facts strongly indicate that Mormonism and at least science as philosophy are basically noncombative.

## Mormon Cosmology

Because Mormonism as a historical phenomenon did not emerge in a vacuum, it is not possible to explore the meaning and relationship of science and religion in the Mormon context without examining the larger dimensions of the scientific and religious communities. Too often studies exploring that relationship deal with the scientific half of the discussion superficially, emphasizing broad scientific themes and facile generalizations at the expense of an analysis of the relevant scientific theories and ideas. Consequently, these studies promote an immature, and even erroneous, sense of the scientific enterprise. Therefore, in part 1 I will explore in detail the philosophic nature of science and its historical development, attempting to examine and then to assess modern science during the years since its emergence in the seventeenth century. Consequently, the reader must be prepared for considerable discussion about the nature, growth, and history of the sciences. Also, I will discuss the relevant religious issues in American life that shaped the broader cultural climate against which Mormonism and science must ultimately be understood. The examination of the scientific and religious background that defined the nature of science will reveal how religion, and specifically Mormonism, has assessed science, one of the most important and powerful intellectual and cultural fields of modern times, and how that science has affected Mormonism as a movement.

Part 2 deals with a variety of general concepts needed to understand the material in the second, which deals thematically and chronologically with cosmological themes specific to Mormonism. Whereas the general approach to Mormon cosmology is historical, issues dealing with the relationship between Mormonism and science and with their scientific, theological, and philosophical foundation are explored in detail. For example, where did various ideas, such as a plurality of worlds and the matter-energy creation of the universe, come from and what gives them their special appeal? To what degree did scientific understanding affect the personal religious views of various Mormon leaders in either preconceptions, motivation, or conclusions? In the case of Mormon expositors of science, is there a unique vision of reality shaped by a Mormon eschatology—the belief in the world's end and the ultimate destiny of humankind? In what sense was science perceived as being true vis-à-vis Mormon religious views? In Mormonism's second century, as it developed intellectual attachments to a variety of scientific ideas, why did some Mormons develop a reactionary response toward science? Because Mormonism incorporated its 'cosmic theology' directly into the very foundations of its theological fabric, is there a natural intersection between these theological ideas and the rise of the science of exobiology, that branch of biology concerned with the search for life outside the earth's environment? Finally, does this mean that Mormonism must now face the dilemmas inherent in a natural theology?

In particular, the various chapters are divided as follows. In part 1, chapter 1 examines historiographical and philosophical issues in science, religion, and Mormonism; this chapter provides the setting for the material in part 2. Chapter 2 assesses developments in modern science, from the seventeenth century to the twentieth, that relate to Mormonism, while chapter 3 explores the conceptual nature of science itself—what constitutes facts, theories, laws, methods, paradigms, and research traditions in science. Although chapter 2 deals with the crucial scientific issues needed to understand the scientific and cosmological themes in Mormonism and chapter 3 overviews essential material on the meaning of science, they may be quickly perused by those who wish to concentrate only on Mormon issues. Neither chapter, however, is meant to offer a comprehensive survey of the history of science and its philosophical foundation.[13] In both chapters I have chosen those historical and philosophical issues characterizing the development and nature of science that in my judgment best reflect the actual workings of the scientific enterprise needed to understand the subject and the approach of this book.

Part 2 is divided into six chapters that explore ideas relating to cosmology, cosmogony, and astronomy. Focusing exclusively on the thought of Joseph Smith, chapter 4 explores the emergence of a unique Mormon cosmology. Chapter 5 examines how this cosmological thinking expanded within the late nineteenth-century church. After Joseph Smith, Orson Pratt was the most important Mormon cosmologist of the nineteenth century. Chapter 6 addresses fully Pratt's cosmology in its scientific, philosophical, and historical setting. Chapter 7 shows that, in the early years of the twentieth century, there emerged a rational scientism within the church hierarchy, mostly as a result of individual

Mormon scientist-authorities B. H. Roberts, James E. Talmage, John A. Widtsoe, and Joseph F. Merrill.

Whereas a positive scientism developed with these church authorities, chapter 8, focusing on evolution as cosmogony, shows that there subsequently emerged within some quarters of the church a reactionary view arguing that modern science is mostly negative. Chapter 9 concludes with a careful examination of several issues in modern cosmology—namely, the creation of the stellar universe and the search for extraterrestrial life—that have the most to say scientifically about Mormon views on cosmology today. Except for the first three chapters, the remaining chapters can be read almost independently of one another.

The section on Mormon cosmology is not a narrative examination of the complete history of the relationship of Mormonism with science. Rather, it explores crucial episodes in that relationship that deal with cosmology both as science and as cosmic vision. Consequently, it does not provide a discussion of all issues from all points of view. Instead, its thematic nature allows discussion of selected but germane issues.

Finally, the reader should be aware of one important caveat. Mormonism has never been dominated by an antiscience view as have some religions. Admittedly, there have been influential Mormon leaders who have been opposed to some aspects of science, but Mormonism as a whole has avoided the negativity to science that is common within many fundamentalist movements. Although reactionary scientific views, as well as those that are, to be blunt, antiscientific and antiintellectual (see chapter 8), are part of the discussion of the relationship of science with religion, in the Mormon context these views are normally on the periphery of that relationship. Those individuals who hold this opinion, however, fail to grasp an essential understanding between Mormonism, science, and scholarship generally. This is not to say that my discussion has prejudicially selected only the proscience views; it does say that this book presents the case that in theological, religious, and historical terms Mormonism and science are *fundamentally* noncombative.

## NOTES

1. Jan Shipps, in her award-winning *Mormonism: The Story of a New Religious Tradition* (Urbana: University of Illinois Press, 1985), has suggested that Mormonism represents the emergence of a new religious tradition characterized by a new cosmology, a new mythopoetic interpretation of self and nature.

2. Throughout this essay I will use the term *science* to refer primarily to the natural and empirical sciences, thus excluding the social sciences and biblical higher criticism.

3. For the best and one of the most recent, compelling examples of a sophisticated approach to Book of Mormon archaeology, see John Sorenson, *An Ancient American Setting for the Book of Mormon* (Salt Lake City: Deseret Book, 1985). Sorenson is in a long tradition of Mormon interest in new-world archaeology, which became institutionalized with the founding of the New World Archaeological Foundation in 1952 by Thomas S. Ferguson, who, along with the Mormon church authority Milton R. Hunter, published *Ancient America and the Book of Mormon* (Oakland, Calif.: Kolob, 1950). Ferguson and Hunter represent the literal-historic tradition in Book of Mormon archaeology at its finest; see Stan Larson, "The Odyssey of Thomas Stuart Ferguson," *Dialogue* 23, no. 1 (Spring 1990), 55–93.

4. In subtle, and sometimes not so subtle, ways, however, technology was fundamentally reshaping the image of humankind and its worldview. See, for example, Leo Marx, *The Machine in the Garden: Technology and the Pastoral Ideal in America* (New York: Oxford University Press, 1967).

5. See Loren Eiseley, *The Firmament of Time* (New York: Atheneum, 1982); and James R. Moore, *The Post-Darwinian Controversies: A Study of the Protestant Struggle to Come to Terms with Darwin in Great Britain and America, 1870–1900* (New York: Cambridge University Press, 1979).

6. See Duane Jeffrey [*sic*], "Seers, Savants and Evolution: The Uncomfortable Interface," *Dialogue* 8, nos. 3/4 (Autumn/Winter 1973), 41–75; Richard Sherlock, "A Turbulent Spectrum: Mormon Reactions to the Darwinist Legacy," *Journal of Mormon History* 5 (1978), 33–59; and Richard Sherlock, "Campus in Crisis," *Sunstone* 4, no. 1 (January/February, 1979), 10–16.

7. Ernst Mayr, "The Nature of the Darwinian Revolution," *Science* 176 (1972), 981–89.

8. In a highly provocative and very suggestive book, Giorgio de Santillana and Hertha von Dechend have argued that all great mythopoetic accounts of human beginnings have one common origin in a cosmology of celestial orientation. See their *Hamlet's Mill* (Boston: David R. Godine, 1977).

9. Although cosmology and Mormonism have cultivated a positive relationship, we should remember that it was cosmology that ostensibly provided the issue of the Catholic church's 1633 trial of Galileo over matters of science; for details of the Galileo episode, see chapter 1.

10. For one recent example of cosmology in the broadest sense, see Karl C. Sandberg, "Knowing Brother Joseph Again: The Book of Abraham and Joseph Smith as Translator," *Dialogue* 22, no. 4 (Winter 1989), 17–37.

11. Jan Shipps, "From Satyr to Saint: American Attitudes toward the Mormons, 1860–1960" (Paper presented at the annual meeting of the Organization of American Historians, Chicago, 1973).

12. Kenneth R. Hardy, "Social Origins of American Scientists and Scholars," *Science* 185 (9 August 1974), 497–506.

13. For detailed studies of the history of science and its philosophical foundations, see the Bibliography.

PART
ONE

Issues in Science and Religion

# 1

## Mormonism and Science

Focusing on the transcendent, religion explains the human drama against the backdrop of cosmic events; in the process it frequently makes claims about the physical universe. To the degree that religion involves itself in the mundane, however, religion and science are seen to clash—often with violent results. Conventional wisdom has it that throughout history science and religion have been, and largely remain, mutually antithetical. No one would deny such well-known episodes as Galileo's conflict with the Roman church or Darwin's threat to nineteenth-century Protestantism. Yet the thesis that religion and science are fundamentally at war cannot be sustained under close historical scrutiny. In this study of the relationship of science to Mormonism, the roots of this misconception will be laid bare. To do so, we must first examine relevant historiographical issues in science and religion and then expose a variety of relevant foundational issues both in traditional Christianity and in the emergence of Mormonism in the nineteenth century.

### Science and Religion

During the last decades of the nineteenth century, two prominent American intellectuals published books dealing with the alleged incompatibility of science and religion: John W. Draper's *History of the Conflict between Religion and Science* (1875) and Andrew D. White's *History of the Warfare of Science with Theology* (1896). Draper, immigrant son of a Methodist minister, and White, president and founder of Cornell University, argued that these two areas of human understanding were fundamentally at odds and essentially always had been. Despite his title, Draper focused on the Roman Catholic church as the arch villain of enlightenment. The Vatican's persecution of scientists, argued Draper, was calculated, deliberate, and designed to smash any ideas, particularly scientific and certain philosophic, that differed with the received view of the church. Draper's study is parochial, however, and his language is emotionally charged and designed to cause anger among its readership inclined toward Catholicism. By contrast, Draper argued that Protestants, with their emphasis on private scriptural interpretation, provided a receptive climate for the emergence of modern science in the sixteenth and seventeenth centuries. Whereas Catholic authoritarianism constantly thwarted science, Protestant pluralism mitigated any tendency toward undue control. Although sharing Draper's anathema for theology, White's study was more broadly conceived. White argued that theology, not religion, had become dogmatic and that this dogmatism regarded the Bible as a scientific text—a mistake that caused an essential conflict between science and religion. The words *conflict* and *warfare* were intentionally used to emphasize a military siege between science and religion, with science clearly on the side of enlightenment, and religion, or at least theology, as the culprit.[1]

The alleged conflict that Draper and White chronicled in their very influential books is, however, fundamentally misdirected. Over time, studies have almost universally shown that the sciences were not at odds with the religious climate of, say, the nineteenth century. The warfare thesis erroneously assumed a set of intellectual categories, such as epistemologies or concepts of nature, that had cast the issue initially in terms of conflict and tension. Recent scholarship views both science and religion as social enterprises in which disputes, when they arose, did so because people possessed different cultural, professional, and political goals, not because there was something inherently antagonistic between science and religion.

To be sure, there were those who believed genuinely that most scientists were agnostic at best, and most likely atheists, and that the effect of theology and religion was to constrain free and open inquiry. In actuality, the conflict Draper and White wrote about was thrashed out in the public press and in the public forum; it was not an attack upon religion by scientists, nor did scientists generally view their work as being antireligious or even areligious. Only with the rise of evolutionary, geological, and anthropological studies, along with the gradual emergence of the 'higher criticism' of the Bible, did some religionists become apologetic, defensive, and in time reactionary. To be sure, the rhetoric generated following such famous conflicts as Galileo's trial in

1633 and the so-called Scopes Monkey Trial of 1925 tended only to fuel a misdirected understanding of science and the intent of the scientific enterprise.[2] But these scientific studies did not preclude the practitioners of science from being personally religious. From the beginning years of the Christian era through the period of this study, the vast majority of those engaged in what we would call science ('natural philosophers' in the parlence of the time) were themselves Christians.

Although the military metaphor has dominated much of the discussion of the relationship between science and religion in the century since Draper and White, subsequently modern scholars have examined the historical matrix from which both science and religion developed and have shown the existence of a very complex and multifaceted relationship between the two. Emphasizing the fact that modern science emerged only in the Latin West, where medieval Catholicism and later Protestantism dominated, they have come to see a more complex and symbiotic relationship between science and religion to account for this singular phenomenon.[3]

During Christianity's first centuries, its attitudes toward natural science, while not wholly different than the major competing philosophies—principally Gnosticism, Neoplatonism, and the mystery religions—urged a greater respect for and study of nature. As David Lindberg, easily one of the most perceptive historians dealing with this issue, has observed, "The church fathers used Greek scientific knowledge in their defense of the faith against heresy and in the elucidation of scripture, thereby preserving and transmitting it during the social and political turmoil of the first millennium of the Christian era."[4] As a result, science became the valued servant of Christianity; they were hardly in opposition.

Perhaps the earliest really fruitful encounter of science with theology, however, occurred in the thirteenth century, following the introduction into the Latin West of the writings of the Greek philosopher Aristotle (384–22 b.c.). Many of Aristotle's scientific concepts, particularly his complementary views of a geocentric cosmology and the physics needed to sustain it, directly supported the church's historical position of an anthropocentric universe. Some of Aristotle's pre-Christian notions, however, were actually in sharp variance with the church's teachings, particularly his view that the world is eternal (contracreationism), that the human soul does not survive its mortal existence, and that God is not omnipotent. The latter view particularly irked the church. After a series of warnings and condemnations issued by various bishops of Paris beginning in 1210, Pope John XXI, concerned with divisive intellectual unrest, instructed the then bishop of Paris, Etienne

Tempier, to investigate. On 8 March 1277, the third anniversary of the death of Thomas Aquinas, the bishop issued a list of 219 propositions. These propositions—some of which were originally directed against Aquinas, who was already revered by some—stated that the teaching of any one of these 219 views would be sufficient grounds for excommunication from the church.

Although the church continued to retain most of Aristotle's scientific views, in their rush to avoid conflict with the church, philosophers (and some theologians) no longer felt constrained to remain faithful to a strict reading of Aristotle. Thus, for example, while Aristotle argued that there could be only one world at the center of the universe, otherwise falling objects would not know which center to fall toward, post-1277 philosophers imagined the possibility of a plurality of worlds in the universe. This approach of imagining possibilities, but not asserting them as realities, came to be known as *secundum imaginationem*—"according to the imagination." In the fourteenth century this theological concern received philosophic sanction from William of Ockham, who argued that God's relationship to the world was completely contingent and could not justify a cause-and-effect relationship between objects seen and causes imagined.

The irony in this intellectual tug between the natural philosophers (i.e., scientists) and the church was that the Condemnation of 1277 freed the philosophers to consider all sorts of other possibilities not originally given in Aristotle's vast scientific and philosophic corpus and thus allowed them to speculate scientifically well beyond any constraints imposed by the writings of Aristotle and the church.[5] The attitude that one could imagine possibilities encouraged the view that science, which at the time was subsumed in philosophy, was ultimately in quest of *logical* consistency (instrumentalism) and not *physical* realities (realism). Partially as a result of this attitude, a full-fledged empiricism never developed, since scientific speculations found confirmation in their agreement with philosophical commentaries and not against the demands imposed by an experimentation with nature directly.

This view began to change most forcefully in the middle years of the sixteenth century, when the astronomer and Catholic canon at Frauenburg in northern Poland, Nicolaus Copernicus (1473–1543), argued that the prevailing view of medieval astronomy, an earth-centered cosmology based largely on the work of the astronomer Ptolemy (fl. 150 A.D.), was fundamentally conceptually flawed. Copernicus, who also held certain Neoplatonic views that emphasized the centrality of the sun, published his radical heliocentric views in his epoch-making

*De Revolutionibus Orbium Celestium* in the year of his death.[6] While other churchmen before Copernicus—such as Bishop Nicole Oresme in the fourteenth century and Cardinal Nicholas of Cusa in the fifteenth century—had freely entertained similar ideas, still others, including another bishop and another cardinal, succeeded in urging Copernicus to publish his book, which was dedicated to Pope Paul III. Although fewer than a dozen astronomers—both Catholic and Protestant—adopted the Copernican hypothesis before the year 1600, heliocentrism became a topic of considerable debate.[7]

Even after Galileo adopted the Copernican view and attempted to force the newer cosmology on traditional biblical understanding, he faced considerable resistance. The arguments he presented in favor of the heliocentric theory—based largely on observations he had made with the telescope, newly built in 1610 and subsequently trained on the heavens—were impressive: he discovered a miniature solar system in the four large satellites of Jupiter; a lunar surface that was rough, mountainous, and generally terrestrial in appearance; solar spots; the "winged" nature of Saturn; an increasingly large number of stars in the Milky Way; and, perhaps most importantly, the changing phases of Venus, which could not be explained using the Ptolemaic-geocentric model of the cosmos. As important as these discoveries were, however, collectively they failed to prove the central thesis of the heliocentric hypothesis: to wit, the annual and diurnal motions of the earth.

This was also the aftermath of the Protestant Reformation and of Catholic reforms initiated by the Council of Trent (1545–63). Religious sensitivities were heightened, and concern with readjusting the physical picture of the cosmos was carefully weighed against questions of scriptural interpretation and traditional scholastic understanding that had proved valid for nearly a millennium. To be sure, the inquisition of the Counter-Reformation reigned supreme for many, but only three scientist-philosophers—the Italians Giordano Bruno, Thomas Campanella, and Galileo—were actually punished for their alleged heresies. Admittedly, Bruno paid with his life in 1600, partially for his cosmological speculations on the doctrine of the plurality of worlds, which countered orthodox thinking, but also for charges of a blatant pantheism and antitrinitarianism and for his involvement in various shady political schemes. In 1599 Campanella was sentenced to life imprisonment (though released in 1626 at the Pope's request) on charges of religious heresy (for dealing in magic and astrology) and political rebellion.

In the end Galileo's 1633 confrontation with the Roman church was caused not by an inherently antagonistic relationship between science and religion but by other problems, including the inconclusive ideas and excessive arrogance of Galileo versus an increasingly theologically conservative church hierarchy. As Jerome Langford has pointed out, Galileo's immediate problem with the church was less a conflict of ideas than a disagreement among the combatants.[8] Still deeper lay an issue between Galileo and the church that did not deal with personalities or with the evidence Galileo had collected to demonstrate his case. Despite the lack of crucial scientific evidence Galileo persisted stubbornly with his views, defending hermeneutic principles violating the church's tradition of scriptural exegesis based on theological, not scientific, interpretation. In the social climate after the Counter-Reformation this was a major mistake. As David Lindberg and Ronald Numbers have argued, it was a struggle between opposing views of biblical interpretation, not a matter of Christianity waging war on science.[9] After all, Galileo and all those involved were devout Catholics. Ultimately, therefore, it was one of interpretation: Galileo espoused a realist conception of science, while for centuries the church had successfully urged an instrumentalism.

The traditional interpretation of the Galileo-church confrontation urged a historical interpretation using such terms as *warfare* and *conflict*. The use of the military metaphor emphasized so forcefully first by Draper and White in the nineteenth century has in the twentieth century simply not been sustained by more clear-minded interpretations of the events. One of the earliest to assess the relationship between science and religion in our times without recourse to the warfare thesis was sociologist Robert K. Merton, who, in his classic study on the emergence of seventeenth-century science, argued that modern science was a product of a Puritan ethos, epitomized in the words "To the glory of God and the good of mankind," that compelled some intellectuals to cultivate an increasing interest in science as a field of study during the seventeenth century. During the period 1640 to 1660 Puritans became the dominant power in England, and in this new era science became an attractive endeavor that promised to reveal the divine creation and improve the standard of living. To demonstrate his case, Merton cited statistics that showed large numbers of individuals entering scientific areas of study. As a way to understand God's world, this Puritan ethos laid the intellectual foundation for the rise of modern science, or at least provided a nurturing environment that fostered the growth of science as we know it today.[10]

Merton's Puritan hypothesis remained a subject of lively discussion, with some emphasizing religion and intellectual ideas as the focus for the emergence of science and others emphasizing technology and the practical application of science as the motive force for change.[11] Although subsequent detailed studies have successfully challenged a number of Merton's conclusions, his view that the social climate that promoted science was fostered in some way by Puritanism can be sustained.[12] Others have shown that, whereas some conflict did exist, the struggle was between groups of professionals, or between competing systems of science for institutional or intellectual control, or within the minds of individuals as a genuine conflict or as a crisis of faith, rather than some all-out struggle between the forces of good and evil. While questions about the significance of Merton's thesis, as well as concerns about definition and methodology, have arisen as central, the debate has clarified both the crucial definition of Puritanism during this period and the relationship between Puritanism and science.[13]

Richard S. Westfall, one of the most influential scholars of the period historians call the Scientific Revolution, has carefully assessed science and its relationship to religion during the seventeenth century. Westfall has argued that many seventeenth-century scientists, such as Isaac Newton and Robert Boyle, understood fully the skepticism and philosophical secularism inherent in the new scientific philosophy that subsequently came to dominate the period.[14] From the beginning years of modern science in the seventeenth century through much of the nineteenth century, at least before Darwin, Protestants held a reasonably positive attitude toward the truthfulness and utility of science. They believed that one could discover the works of God in his universe and that those works would complement His words in the Holy Scriptures. This belief evolved into what became known as 'natural theology'which has since become a dominant theme in the Anglo-American world.

The classic view of natural theology was given most forcefully by the English theologian William Paley (1743–1805) in his *Natural Theology*. The author of three books, one on moral philosophy and two defending Christian beliefs, Paley relied on a variety of natural phenomena to establish the existence of God. Using a metaphor that has since come to epitomize the argument by 'design,' Paley stated that if one were to find a watch on the ground one would be impressed with its mechanism and generally its laws of operation, and would rightly infer the existence of its architect. Drawing upon human, animal, and insect anatomy and physiology, Paley argued that nature provides overwhelming evidence of intricate mechanisms. Since mechanisms imply a designer and since there is uniformity and order in nature, there must be a single (divine) intelligence at work.[15] By the time natural theology had run its course, roughly mid-nineteenth century, it had

changed from observing God *in* nature to observing nature and de-
claring its motions the will of God.

Ironically, in defending Christianity through the tenets of natural
theology, Christian scientists of the seventeenth and eighteenth centuries prepared the ground for the deists of the Enlightenment. And
in time a radically different worldview emerged from the writings of
the eighteenth-century deists: the mechanical universe governed by
immutable natural laws, the wholly transcendent God completely removed and separated from his creation, the moral law taking the place
of spiritual worship, the rational man able to discover true religion
without special revelation. Once the reverence for Christianity was
removed, deism, the religion of reason, steps full grown from their
writings. With these developments in the eighteenth century, natural
religion (or deism) and natural theology separated and became radically
different enterprises. Even with these developments, however, science
and religion were not generally at odds. During this period, for instance,
very few Americans became deists or skeptics, and if religious orthodoxy declined, as is commonly claimed, the decline resulted more from
prosperity, complacency, and sectarian bickering than from science and
its secular influence.[16]

With the emergence of deism came its intellectual offspring naturalism, the view that only science properly conceived allows for an
investigation and explication of nature's laws. Generally this was not
an antireligious ideology but the reaffirmation that nature is governed
by regularity, not by chance or random causes. Admittedly, there were
a few who developed a full-fledged antireligious materialism, such as
the scientific materialists in nineteenth-century Germany. But most
other scientific movements whose heritage can be directly traced to
naturalism—such as vitalists, French materialists, and German reduc-
tionists—were less interested in promoting religious conflict and more
engaged in developing an understanding of nature that simply avoided
religious concerns.[17]

Ronald Numbers, one of the most perceptive contemporary scholars
of science and religion, has rightly argued that this eighteenth- and
early-nineteenth-century naturalism prepared the ground for evolutionary ideas later in the nineteenth century.[18] Since both Draper and
White wrote their books in the post-Darwinian climate of late nineteenth-century America, they suffered from an excessive polarization
of perceived hostilities due to the rhetoric and polemics generated in
the Darwinian aftermath. Before Darwin published his *Origin of Species*
in 1859, however, most Anglo-American scientists were, if not orthodox Christians, predisposed to a natural theology that supported the

religious view that nature directly reflected the design, harmony, and order of a benevolent divine architect. Admittedly, Darwin's emphasis exclusively on natural causes, regardless of its religious implications, eventually may have offended many devout (and traditional) Christians. As Ernst Mayr, one of the foremost evolutionists in America today, has argued, in the end the Darwinian revolution replaced 'special creation', 'essentialism', 'catastrophism', 'anthropocentrism', and the 'young earth' with a naturalism that affirmed little if anything about divine presence.[19]

Recently, Peter Bowler has asserted that this traditional interpretation of a revolutionary Darwinism has been exaggerated and even misstated. By reexamining developments in nineteenth-century human evolution, evolutionary social thought, and non-Darwinian evolution, Bowler claims that Darwin's evolutionary theory was not the central feature of nineteenth-century evolutionism and therefore not a primary threat to theology. Rather it was only a *catalyst* in the transition to an evolutionary viewpoint that came to reject key features of theological creationism, particularly the distinction between humans and the rest of nature. In contrast to Mayr, Darwin succeeded in preserving a teleological view of nature. Consequently, Bowler argues that the interpretation of a nineteenth-century Darwinian revolution is a myth.[20] Not until the synthesis of genetic theory and evolution after the turn of the century could one claim a real Darwinian revolution—and then only in the scientific sense.

Still, Darwin was engaged not in theological discourse but in affirming the regularity of the cosmos and its dominance by natural laws. In his devastating critique of the military metaphor, James Moore has shown that the dispute between the conservative evangelical community and Darwinism resulted more from a disagreement between professional geologists and professional biblical scholars on the one hand and amateur exegetes and scriptural geologists on the other rather than from some inherent antagonism between science and religion.[21] Consequently, while the Darwinian debates created genuine conflict, it was not between scientists and theologians but, as Moore argues, a "conflict of minds steeped in Christian tradition with the ideas and implications of Darwinism."[22] Except for very conservative (and mostly literalistic) religious groups, such as Methodists, Baptists, and Lutherans, Moore has also convincingly shown that most American Protestants, principally Congregationalists, Unitarians, and Presbyterians (who alternated in their opinions, however), made their peace with Darwinism by the turn of the century.[23]

Some have argued, however, that within the scientific community of the late nineteenth century deep hostility remained. In his sophisticated and sometimes controversial study, Neal Gillespie has insisted that, although scientists and clerics made peace with Darwin, conflict nevertheless existed between proponents of an older theologically grounded science and those of the newer positivistic science that had excised the supernatural.[24]

Since the founding of modern science in the seventeenth century, the majority of scientists simply avoided religious questions and theological controversy by emphasizing that the legitimate role of science assumed that natural law governed the cosmos. Some scholars have deemphasized an alleged conflict between science and religion and have argued that science itself, interpreted as a *human* enterprise, is often shaped along religious lines.[25] Still others have found the military tension less in some sort of inherent philosophical relationship between science and religion, and have seen the conflict as focused in the personalities involved.

Only during the last few decades has it become eminently clear that the military metaphor is far too simplistic and, in most cases of alleged conflict, utterly misleading and totally mistaken. As human enterprises, not only are science, religion, and their relationship extremely complex, as defined by unique ideas which themselves may be in tension, but the mere fact that science and religion also have institutional, social, and even economic dimensions enriches and complicates that relationship. Both science and Christianity have evolved within the same cultural climate, and in the process they have profoundly shaped that climate. Consequently, argue Lindberg and Numbers, "in the future we must not ask 'Who was the aggressor?' but 'How were Christianity and science affected by their encounter?' "[26]

### Science and Mormonism

Because Mormons rejected traditional religion and theology while only tacitly endorsing science, they have stood somewhat outside the intellectual dimensions of normative science and religion. Mormonism's basic claim is that its doctrines constitute a restoration of primitive Christianity—a regeneration of the pristine gospel and the unadulterated principles existing prior to an apostasy that occurred during the early centuries of the Christian era. Since science in its modern form is the product of orthodox Christian culture, the argument goes, it too has become infested with false and misleading principles. Therefore, Mormonism continually emphasized that the study of nature, properly understood through Mormon theology, would lead to "true" science. The writings of nineteenth-century, and of even many twentieth-century, Mormons, as well as of many Christians generally, emphasized the analogy that secular science is to apostate Christianity as "true" science is to the restored gospel of Mormonism. After all, God is the author of both the revealed word (scriptures) as well as of nature itself. And because God is trustworthy, both his words and his works must be entirely compatible.

Although many scholars took note of the ideas of Draper and White and used them profitably at times, both Draper and White wrote, at least partially, for public consumption—and much of the public was influenced. The Mormons were no different. They took great interest in Draper and White, and basically agreed with them. Whereas Draper and White, in the Mormon view, were dealing with uninspired science and apostate religion, they both argued that there is no conflict between true science and true religion.[27] This point was not lost on the Mormon mind, and virtually every Mormon writer argued that there would simply be no conflict between divine, or true, science and revealed religion, or Mormonism. The rhetoric was heavy and editorialized widely in the Mormon press. Referring to the alleged conflict between religion and science, one Mormon editorial declared: "No truth is evil, for all truth is divine. . . . In fact, true science is as divine in its sphere as religion."[28] Elsewhere, "the true theologian can no more antagonize science than he can deny revelation; he can no more ignore the results of true philosophical investigation than he can discard the authentic statements of the Scriptures."[29] In 1882 George Reynolds, secretary to the First Presidency, expressed the view common among Mormons that "the Bible . . . was not written to teach men science by scientific methods."[30] In 1903 and 1904 scientist and later Mormon apostle John A. Widtsoe published a series of articles in the *Improvement Era* under the title *Joseph Smith as Scientist*. Although uncritical in their overall approach, these articles, the title of which sums up the Mormon belief that science is divinely inspired, were later reprinted in revised form under church supervision in 1908.[31]

This tradition of science divinely understood, however, can be traced at least as far back as St. Augustine in the fourth century. In the seventeenth century it was centrally embodied in the development of natural theology. For example, Isaac Newton held that monotheism, what he considered to be the pristine religion of Christianity, was intimately related to the true system of the world, heliocentrism. The same view had been given wide currency by nineteenth-century Adventists, while today it is identical to the views of religious creationists.

The question of concern for Mormonism is how it managed to accom-modate itself to a rapidly growing scientific ethos that increasingly dominated the cultural climate of Western thinking—even on the Amer-ican frontier. The answer partially lies in the attitude toward nature conceived as the handiwork of God. Mormons adopted a traditional view toward natural law most fully developed by the philosophers who first laid the foundations of modern science in the seventeenth century. Equally, in order to protect their nascent religion, Mormons also rejected both a severe naturalism, which in the hands of some led to religious skepticism, and a strict natural theology.

Even as Mormons ostensibly recognized the distinction between true science divinely inspired and the actual science of the day, they rapidly made use of modern science in their efforts toward theological expli-cation.[32] Although Joseph Smith introduced numerous ideas that have found scientific parallels, the earliest sustained use of a scientism to support various aspects of Mormon theology was given by Parley P. Pratt and his brother Orson. Beginning in 1840 Parley, the first Mormon authority to write religious tracts reflecting restorationist themes, used cosmological and eschatological speculations widely. While Parley en-gaged in poetic speculation, Orson, beginning around 1845, felt more comfortable in developing the new cosmological ideas philosophically and scientifically.[33] Extensive development of cosmological themes first emerged in Parley P. Pratt's *Key to the Science of Theology* (1855), the first book-length study exploring the scientific dimensions of Mormon theology. These developments continued in various articles and con-ference addresses later found in church-related publications, such as Orson Pratt's work in the *Millennial Star* and church authority talks in the *Journal of Discourses*.

Adhering to the general validity of the warfare thesis of Draper and White and adopting the view of true science in light of restorationist thinking, Mormons urged the view that science properly conceived was good and useful—something to both cultivate and emulate. The first sustained studies by Mormons to explore the relationship of science to religion appeared toward the end of the nineteenth century. John H. Ward's *Gospel Philosophy: Showing the Absurdities of Infidelity and the Harmony of the Gospel with Science and History* (1884) explicitly adopted the warfare thesis, arguing that Mormonism, as revealed re-ligion, was fully compatible with God's science and that secular science was inherently corrupt. The earliest sustained use of evolution and a scientism broadly conceived to support Mormon theology was pub-lished soon afterward in 1895 by Mormon Nels L. Nelson in a series of six articles, "Theosophy and Mormonism."[34] In 1904 these appeared in expanded form in his *Scientific Aspects of Mormonism*. Although Nelson applied scholarly ideas widely, he was particularly fond of using ideas of progress and evolution to argue for the spiritual and physical development of God as well as humans. In his view evolution did not act in a blind, nondeterminative way; rather, spiritual forces via God directed physical processes, such as the laws of physics, biology, and chemistry.

Even so, as a religion Mormonism has accepted many, if not most, scientific ideas. Beginning during the early decades of the twentieth century, after B. H. Roberts became a member of the First Quorum of the Seventy and James E. Talmage, John A. Widtsoe, Joseph F. Merrill, and Richard R. Lyman became members of the Quorum of the Twelve, they produced a large number of church materials that relied heavily on contemporary scholarly thinking and were based on the premise that science—properly understood—is good and that Mormonism and science are mutually supportive.[35] For example, some of their more influential writings include Widtsoe's *Joseph Smith as Scientist* (1908), which was later reissued (1920–21) in revised form as a manual for the Mutual program, and his *Science and the Gospel* (1908–9), which admittedly were both published before he became an apostle in 1921; his *How Science Contributes to Religion* (1927), used in the Mutual program along with *Heroes of Science* (1926–27), which Widtsoe coauthored with Franklin S. Harris and N. J. Butts; and his *In Search of Truth: Comments on the Gospel and Modern Thought* (1930); Roberts's *Joseph Smith: The Prophet-Teacher* (1908), his five-year series *The Seventy's Course in Theology* (1908–12), and his unpublished magnum opus *The Truth, The Way, The Life: An Elementary Treatise on Theology* (1928); Talmage's influential church-published pamphlet on geology and evolution *The Earth and Man* (1930); and Merrill's *The Truth-Seeker and Mormonism* (1946), a collection of essays originally given as Sunday evening radio talks. These authorities, as well as an increasingly larger group of Mormon scientists and scholars, produced numerous articles for church-related publications and gave many sermons and fireside talks in which they discussed science and its compatibility with Mormon theology.[36] Even those who became somewhat more cautious about the implications and truth claims of science, such as Widtsoe in his later years, portrayed science in a positive light.

A number of general authorities, including the scientifically inclined Roberts, Talmage, Widtsoe, Merrill, and Lyman, adopted the view for an old-earth. Except for possibly Roberts, however, they mostly rejected the Darwinian interpretation of evolution. During the early years of the twentieth century other authorities, principally Joseph Fielding

Smith, militantly rejected both evolution and the old-earth view.[37] Still, there were other believing Mormons who were not so reticent. William H. Chamberlin's *Essay on Nature* (1915) and Frederick J. Pack's *Science and Belief in God* (1924) were both strong defenses of evolution and impressive attempts to reconcile evolution with religion.[38]

Although Mormons—authorities, scientists, and layman alike—frequently accepted uncritically much of the rhetoric of the warfare thesis, in practice they were extremely receptive to most of mainstream science. For them, the view that science is a quest for knowledge was a divinely inspired injunction, and that knowledge properly understood would eventually be to their eternal betterment.

## Mormonism and Philosophic Realism

Within the Mormon canon it has often been construed from statements such as "the glory of God is intelligence" (D&C 93:36) and "if a person gains more knowledge and intelligence in this life . . . he will have so much the advantage in the world to come" (D&C 130:19) that perhaps God's science is only more refined or advanced than ours. Except for Mormonism, few Christian religions, contemporary or otherwise, have so adamantly asserted the unity of faith and reason. Such claims as "all blessings are predicated on laws properly understood" (D&C 130:20) and "spirit is merely a more refined form of matter" (D&C 130:20) suggest that, despite significant shades of materialism in these views, the Mormon religion embraces a full-fledged philosophic realism. Both metaphysically and epistemologically, Mormonism asserts the independent existence of universal attributes such as justice, mercy, goodness, and law as well as a direct correspondence between mind and being, between the perceiver and the thing being perceived.

This epistemological realism makes several claims concerning reality and humans. First, it asserts the existence of an objective reality apart from whether humans exist or not. Second, it asserts that humans can perceive this objective reality and that they can construct a clear understanding of it. Finally, it asserts what is real in terms of what is knowable. In this view humans engage science as an enterprise that provides a real understanding of the world, of objective reality. Thus scientific claims are not merely mental constructs or human inventions that are simply instruments of understanding. Nor, in this view, is science merely model building. Rather, science genuinely provides insight into the workings of the natural world. This does not suggest that all scientific claims are truth claims but that science is an intellectual endeavor that allows humans the capacity eventually to uncover an objective reality.[39]

Consequently, various statements found in the canonized writings of the restoration have urged upon historical Mormonism a remarkably positive attitude toward science and the natural order of things.[40] This view makes it virtually impossible to defend the thesis that science and religion are radically different enterprises—science, on the one hand, seeking truth and religion, on the other, providing meaning. Historically, Mormonism has *opposed* relegating religion only to those activities concerned with God, values, and ultimates where the emphasis is prescriptive and extraphysical and defining science as non-theistic, nonteleological, nonethical, nonmetaphysical, descriptive, and terrestrial. Given the uncritical blending of science and religion and the religious injunction toward a realist vision of reality, Mormonism actively encourages an epistemological pluralism that allows for an understanding of the world from both religious and scientific perspectives.

## Mormonism and Natural Theology

The basic premise of natural theology is that nature contains clear, compelling evidence of God's existence and perfection. Moreover, as a study in rational religion, natural theology asserts that the Christian God created a universe in which laws, design, purpose, and harmony are paramount, and the scientist, usually a Christian, could find justification for his religious convictions in his scientific studies.

While deism was a significant, though not dominant, influence in early nineteenth-century America, natural theology continued to have an important effect on many scientists. Mormons, however, never fully accepted Protestant natural theology because their understanding of God and his world was informed first by revelation (for instance, the Book of Mormon, the Doctrine and Covenants, and the Pearl of Great Price, which were given by revelation) and not by science and a study of natural phenomena. In the final analysis the evidences for God did not require a study of nature. Still, properly undertaken, a study of nature would reveal the divine presence. Generally, Mormons, as with Christian natural theology, revered the works of nature. Thus, as one writer expressed it: "The pressing need of the age is a system of religion that can recognize, at the same time, the truths of demonstrated science and the doctrines found in the pages of sacred writ, and can show that perfect harmony exists between the works and words of the Creator; a religion that will reach both the head and the heart—that is, will satisfy both the intellect and the conscience."[41] Consequently, not only was God's word a testament of his continuing interests in human affairs but also his *works* give abundant evidence of the nature, power, and majesty of the divine presence. This statement, written in 1898, would have been acceptable in 1830 or even late in this century.

The Mormon attitude toward science historically has been genuinely positive, precisely because science, properly interpreted, lends credence to the Mormon view of the universe. "Since the establishment of the Church of Jesus Christ of Latter-day Saints," editorialized the Mormon church's *Deseret News* in 1899, "scientific research has furnished many proofs of the truth of 'Mormonism,' and they are still accumulating."[42] Published under the title "Science Proves 'Mormonism,' " this editorial reflects views occasionally appearing in today's church publications, such as the *Church News*. The selective use of scientific findings and theories to support the Mormon position are numerous, but in the final analysis they have nearly always been irrelevant.

### Mormonism and Natural Law

The central premise of classical natural law is that all things have a purpose and that humans, being rational, can uncover that purpose by conforming their rational understanding to nature. With the Reformation of the sixteenth century this view of natural law became more restrictive. In the relationship of natural law to God, the challenge to orthodox (Christian) theology was the degree to which God acts independently of his creation. If God has a necessary, dependent relationship to his creation, then the role of natural laws is compromised at best and irrelevant at worst. In order to protect the glory of God and to avoid making God's actions dependent on the actions of his created beings, the seventeenth-century Protestant reformers affirmed the idea of radical sovereignty, which excluded the active contribution of lesser beings to his work, against the medieval idea of accommodating sovereignty.[43] Consequently, the very stuff of nature itself was deemed utterly passive. Nature was neither active, animate, nor psychic; in short, no active principles, whether sympathies or antipathies, attractions or repulsions, were allowed whatsoever. In order to guarantee the operation of his creation, God therefore imposed laws onto the ontology of nature.

Within Mormonism this transition from an accommodating to a more severe (but never radical) sovereignty occurred over the course of the nineteenth century. Simultaneously, the view that humans possessed a premortal state increasingly opposed the idea that humans (and all of God's creations) are only in a contingent, rather than a necessary, relationship to God. The source of these opposing views—that is, the existence of a sovereign God whose relationship to humans was necessary—and the various roles played by God, humans, and natural law were complicated by competing intellectual traditions. During the 1820s and '30s the Mormon milieu incorporated religious sensibilities that favored a Christian literalism, a return to primitivist notions and Enlightenment thinking.

Although acknowledging a God sovereign over human events—that is, a God who at one extreme directly controls human and natural affairs but does not have a reciprocal relation with his creations—Mormons also recognized the dual role of human free will and the causal effect of natural law as operative in a benign universe. Early Mormons gave greater credence to the role of God in influencing the events of humans than Mormon apologists and interpreters of the Mormon experience during the later years of the nineteenth and twentieth centuries. Whereas an active role was assigned to God, thus protecting the prophetic and revelatory in Mormonism, the Mormon canon also increasingly affirmed the agency of humans and the role of natural law.

Over the course of its written history, there has existed in Mormonism a continuum of views on the role of causation in both human and universal affairs from an extreme supernaturalism on the one hand to a complete naturalism on the other. Early devout interpreters chose an extreme form of supernaturalism in order to assure the affirmation implied in such human events as the First Vision, the coming forth of the Book of Mormon, and the (re)establishment of the institutional church. Toward the end of the century, Mormon apologists, although acknowledging the role of the supernatural, increasingly emphasized the role of humans and other naturalistic influences in the rise of Mormonism. In more recent times the new Mormon history has empha sized natural causation and human agency, sometimes at the expense of the supernatural. In any case, this conscious intellectual drift toward naturalism reflects larger events in the American intellectual tradition.[44]

Although very early in its history Mormonism defined God as material and personal, and therefore operating in both time and space, it also saw God as less subject to natural law. Mormons frequently viewed God's role in human affairs, and particularly in the events leading to the restoration, as primarily vertical. But in most other events in the *physical* universe his role was passive and bound horizontally to the causative nature of universal law. Normally the only exceptions were crucial events directly connected to the course of human history. When dealing with events removed directly from its prophetic foundations, Mormonism generally expressed attitudes supporting an understanding of the world and of science as both lawful and useful.

On the issue of natural law Protestant evangelicals were more deeply divided. Those who espoused natural theology as a worthy endeavor recognized the importance of natural law, at least until roughly mid-nineteenth century, when geological and evolutionary studies, with their heavy emphasis on natural law—and only natural law—made it increasingly difficult to argue for the divine origin of humans. After Darwin published his *Origin of Species* (1859) and his *Descent of Man* (1871), Christians increasingly separated into two intellectually and theologically opposing groups—one becoming more fundamentalist and literalist, the other increasingly secular and humanist—a division, however, which has become mitigated in recent years.[45]

Mormons agreed with those espousing natural theology on the issue of natural law, never relinquishing, before or after Darwin, the importance of this view. Evidences for this somewhat unique Mormon view abound throughout the history of church publications and many commentaries published by, and in defense of, the church. Even after Darwin had published his evolutionary theory that resulted, in Loren Eiseley's words, in making the world *natural* and virtually deifying the idea of natural law, Mormons continued strongly to support a naturalistic view of the physical world.[46] For instance, in 1882, twenty years after Darwin, George Reynolds argued in reference to non-Mormon Christians: "The believer claims God's direct control in mundane affairs; the skeptic maintains the rule of universal law, independent of a personal Deity or other controlling personage. Both are right; God rules indeed, but he rules by law. His laws are universal, eternal, undeviating, and inexorable."[47]

The historical relationship between Mormonism and science was argued succinctly by Dr. Frederick James Pack, professor of psychology and philosophy at Brigham Young College, in an essay originally printed in the student *Crimson* and then published in the church's *Improvement Era* of 1908 because the editors of the *Era* felt that "its merits deserve . . . a wider circulation."[48] "Christian civilization," wrote Pack, "is today divided into two powerfully opposing factions, the supernaturalists and the naturalists. The one believes in an overriding and interposing being, while the other recognizes in nature adequate inspiration without the intervention of the supernatural. The one accounts for the creation of the universe as the result of divine edict; the other can see nothing but obedience to natural law. Thus the warfare between religion and science is being waged. . . . The unending strife between science and religion is very largely the result of this artificial classification of God's laws into the natural and the supernatural. . . . *The term supernatural should become obsolete at once.*" Consistent with this "authorized" Mormon position, Pack suggested that "laws should be classified as *known* and *unknown* and *all as natural*. . . . It is high time that all of God's laws are recognized as natural. He made 'heaven and earth, the sea, and all that in them is'; and therefore the laws that govern these things must be recognized as pertaining to him. *Natural laws are God's laws.*" "No warfare," concluded Pack, "exists between 'Mormonism' and true science."[49]

Most other Mormon thinkers before and after Pack continued to argue precisely in this way because of two important Mormon developments. First, although Mormons frequently employed a supernaturalism in interpreting the *human* events of the restoration, a Mormon naturalism is found in the restorationist scriptures that sanctions views dealing with 'intelligence', 'knowledge', 'laws', 'free will', and so forth. Second, Joseph Smith had reinterpreted a number of critical terms that supported a naturalism, introducing some novel ideas that directly contradicted the traditional Christian view of miracles, supernaturalism, and creationism. Specifically, Joseph Smith redefined the terms 'spirit', which he interpreted as a "material substance, only more refined," and 'creation', which, in his understanding, meant "to organize from pre-existing materials" rather than the emergence of something *ex nihilo*. Augmented by various interpreters of the Mormon canon, these ideas had a crucial affect on the development of a Mormon natural law that was metaphysically much closer to developments in science than to non-Mormon religious traditions.

### Mormonism and Modern Science

Historically, scientific ideas have found their expression in numerous Mormon publications that reveal that Mormons were and are deeply interested in a variety of scientific topics.[50] Since Mormonism has carefully and deliberately articulated a cosmology and cosmogony that has given shape to much of its theology, it is possible to expose Mormonism's philosophical relationship with modern science. As organized in subsequent chapters, the material here corresponds broadly to Mormonism's worldview, to its unique cosmology as it developed historically. Although the issues are grounded in their cultural context, they are also treated philosophically, so that the cosmology that has come to define Mormonism and its relationship to science is clearly evident.

As we explore this relationship, we will see that several complementary views have come to dominate Mormonism's understanding
of science. While Mormonism encouraged a positive scientism, its insistence on a philosophical realism, its adoption of the idea of natural
law, and its view that true science and true religion will never conflict
have all encouraged both sympathies with a natural theology and an
epistemological realism often urging an excessive and uncritical acceptance of science tempered only by occasional prophetic statements.
Failure to understand this results in a complete misunderstanding of
the relationship of Mormonism to modern science. In the broader context of the relationship of religion generally to modern science it is
imperative that one grasp fully that science is as much a human enterprise as religion. Mormonism, which obviously shares similar social
dimensions, is also ultimately grounded in a set of conceptual and
methodological assumptions that has profoundly shaped its understanding of science. In addition to the philosophical developments discussed above—realism, natural theology, and natural law—the relationship of science and Mormonism has been tempered by various
theological developments within Mormonism—such as metaphysical
pluralism, eternalism, and developmental (process) theology—that will
be exposed as they relate to the discussion.

NOTES

1. At the time of their publication, Draper received twenty-seven critical
reviews, including one Mormon review, while of White's twenty-eight reviews,
the eleven that were religious in tone included eight from Protestants, two
from Catholics, and one from a Mormon source. See Donald Fleming, *John
William Draper and the Religion of Science* (New York: Octagon Press, 1972),
and David C. Lindberg and Ronald L. Numbers, "Beyond War and Peace: A
Reappraisal of the Encounter between Christianity and Science," *Church History* 55 (1986), 338–54, which reappraises White's warfare thesis in light of a
reassessment of the relationship of science and religion.

2. Jerome Langford, *Galileo, Science and the Church* (Ann Arbor: University of Michigan Press, 1971); Ray Ginger, *Six Days or Forever: "Tennessee
v. John Thomas Scopes"* (Boston: Beacon Hill, 1958); and Paul M. Waggoner,
"The Historiography of the Scopes Trial: A Critical Re-evaluation," *Trinity
Journal* 5 (1984), 155–74.

3. Pierre Duhem, *Medieval Cosmology*, trans. R. Ariew (Chicago: University of Chicago Press, 1985); Eugene M. Klaaren, *Religious Origins of Modern
Science* (Grand Rapids: Eerdman's, 1977); Stanley Jaki, *Origin of Science and
the Science of Its Origins* (Edinburgh: Scottish Academic Press, 1978); and idem,
*The Road of Science and the Ways to God* (Chicago: University of Chicago Press,
1978).

4. Lindberg and Numbers, "Beyond War and Peace," 342.

5. For accounts of the effects of the condemnation, see Edward Grant, "The Condemnation of 1277, God's Absolute Power, and Physical Thought in the Late Middle Ages," *Viator* 10 (1979), 211–44, reprinted in Edward Grant, *Studies in Medieval Science and Natural Philosophy* (London: Variorum Reprints, 1981), article 13. Also see Edward Grant, *The Physical Sciences in the Middle Ages* (New York: John Wiley, 1971); David C. Lindberg, "Science in the Early Church," in *God and Nature: Historical Essays on the Encounter of Christianity and Science*, ed. David Lindberg and Ronald Numbers (Berkeley: University of California Press, 1986), 19–48; and idem, "Science and Theology in the Middle Ages," in ibid., 49–75.

6. Thomas S. Kuhn, *The Copernican Revolution* (Cambridge, Mass.: Harvard University Press, 1957).

7. Robert S. Westman, "The Copernicans and the Churches," in *God and Nature*, ed. Lindberg and Numbers, 76–113.

8. Langford, *Galileo, Science and the Church*, passim. See also William R. Shea, "Galileo and the Church," in *God and Nature*, ed. Lindberg and Numbers, 114–35; Olaf Pedersen, "Galileo and the Council of Trent: The Galileo Affair Revisited," *Journal for the History of Astronomy* 14 (1983), 28–29. On Galileo's 1616 confrontation with the church, see Richard S. Westfall, *Essays on the Trial of Galileo*, Studi Galileiani, no. 5 (Vatican City: Vatican Observatory Publications, 1989).

9. Lindberg and Numbers, "Beyond War and Peace," 346–47.

10. Robert K. Merton, *Science, Technology, and Society in Seventeenth-Century England* (New York: Fertig, 1970), originally published in 1938 in *Osiris: Studies in the History and Philosophy of Science, and on the History of Learning and Culture*, vol. 4, pt. 2. For an essential contemporary assessment of this view, see I. Bernard Cohen, ed., *Puritanism and the Rise of Modern Science: The Merton Thesis* (New Brunswick: Rutgers University Press, 1990).

11. Richard F. Jones, *Ancients and Moderns: A Study of the Rise of the Scientific Movement in Seventeenth-Century England* (St. Louis: Washington University Press, 1961); Christopher Hill, *Intellectual Origins of the English Revolution* (New York: Clarendon Press, 1966); and R. Hooykaas, *Religion and the Rise of Modern Science* (Edinburgh: Scottish Academic Press, 1972).

12. See R. Hooykaas, "Science and Reformation," *Journal of World History* 3, no. 1 (1956), 109–41; and David Kubrin, "Newton and the Cyclical Cosmos: Providence and the Mechanical Philosophy," *Journal of the History of Ideas* 28 (1967), 325–46.

13. Charles Webster, *The Great Instauration: Science, Medicine, and Reform, 1626–60* (London: Duckworth, 1975).

14. Richard S. Westfall, *Science and Religion in Seventeenth-Century England* (Ann Arbor: University of Michigan Press, 1973).

15. William Paley, *Natural Theology*, ed. Frederick Ferre (Indianapolis: Bobbs-Merrill, 1963).

16. Henry F. May, "The Decline of Providence?" in his *Ideas, Faiths, and Feelings: Essays on American Intellectual and Religious History, 1952–1982* (New

York: Oxford University Press, 1983), 130–46; and Herbert Leventhal, *In the Shadow of the Enlightenment: Occultism and Renaissance Science in Eighteenth-Century America* (New York: New York University Press, 1976).

17. See Roger Hahn, "Laplace and the Mechanistic Universe," in *God and Nature*, ed. Lindberg and Numbers, 256–76; Jacques Roger, "The Mechanistic Conception of Life," in ibid., 277–95; Frederick Gregory, *Scientific Materialists in Nineteenth-Century Germany* (New York: D. Reidel Press, 1977), passim; and E. Robert Paul, "German Academic Science and the Mandarin Ethos," *British Journal for the History of Science* 24 no. 1 (1984), 1–29.

18. Ronald Numbers, *Creation by Natural Law: Laplace's Nebular Hypothesis in American Thought* (Seattle: University of Washington Press, 1977).

19. Ernst Mayr, "The Darwinian Revolution," *Science* 176 (1972), 981–89.

20. Peter J. Bowler, *The Non-Darwinian Revolution: Reinterpreting a Historical Myth* (Baltimore: Johns Hopkins University Press, 1988).

21. James R. Moore, *The Post-Darwinian Controversies: A Study of the Protestant Struggle to Come to Terms with Darwin in Great Britain and America, 1870–1900* (New York: Cambridge University Press, 1979), 19–122. Also see Jon H. Roberts, *Darwinism and the Divine in America: Protestant Intellectuals and Organic Evolution, 1859–1900* (Madison: University of Wisconsin Press, 1988).

22. Moore, *The Post-Darwinian Controversies*, 102–3.

23. For the best exposition of the relationship of science to religion in America from the colonial period to modern times, see Ronald L. Numbers, "Science and Religion," *Osiris* 2d ser., 1 (1985), 59–80.

24. Neal C. Gillespie, *Charles Darwin and the Problem of Creation* (Chicago: University of Chicago Press, 1979).

25. Charles C. Gillispie, *Genesis and Geology: The Impact of Scientific Discussion upon Religious Beliefs in the Decades before Darwin* (New York: Harper and Row, 1959); and Herbert Hovenkamp, *Science and Religion in America, 1800–1860* (Philadelphia: University of Pennsylvania Press, 1978).

26. Lindberg and Numbers, "Beyond War and Peace," 354.

27. All of the twenty-seven critical reviews of Draper, including the LDS review, agreed with him that there would be no conflict between true science and true religion.

28. Anonymous, *Millennial Star*, 37 (26 July 1875): 464–65. References in the primary Mormon literature to ideas on science, such as natural law, evolution, and the plurality of worlds, are extensive. Rather than list countless possibilities, therefore, I have chosen representative samples.

29. *Deseret News* (1899) reprinted in *Millennial Star* 61 (23 November 1899), 751.

30. George Reynolds, "The Conflict between Science and Superstition," *Millennial Star* 44 (17 April 1882), 253–55, 270–71, esp. 254.

31. John A. Widtsoe, *Joseph Smith as Scientist* (Salt Lake City: YMMIA Publications, 1908).

32. The tangible fruits of science were also welcomed. For instance, Parley P. Pratt, writing in 1840, noted that with regard to the means needed

for the gathering and restoration of Israel, "God has inspired men . . . to invent all these useful arts and improvements." See Parley P. Pratt, "The Millennial," *Millennial Star* 1 (August 1840), 74. Others suggested that "true science and true religion . . . work under the same Guiding Hand, for the development, progress and eternal exaltation of mankind in time and eternity." See *Millennial Star* 37 (26 July 1875), 466.

33. Parley P. Pratt, *The Millennium, and Other Poems* (1840); Orson Pratt, "Angels," *New York Messenger* (September 1845), later reprinted as "Mormon Philosophy," *Millennial Star* 7 (1846), 30–31.

34. Nels L. Nelson, "Theosophy and Mormonism," *Contributor* 16, nos. 7–12 (May–October 1895), 425–31, 482–91, 562–68, 617–25, 698–705, 729–39.

35. Many of the science-related articles published by Widtsoe and Merrill, along with articles by Henry Eyring, Carl Christensen, Harvey Fletcher, Frederick Pack, and Franklin S. Harris, all trained scientists and prominent Mormons, were later compiled into the book *Science and Your Faith in God*, comp. Paul Green (Salt Lake City: Bookcraft, 1958).

36. The primary literature substantiating this view is extremely large, but in the twentieth century nearly every issue of the *Improvement Era*, as well as the Conference Reports and later the *Church News* and other church-related publications, carried articles on science and Mormonism. See Robert L. Miller, "Science/Mormonism Bibliography, 1901–1989."

37. While somewhat outside the bounds of this study, for a discussion of the antievolution, anti-old-Earth reaction within Mormonism, see chapter 8.

38. For an analysis of these works, see Richard Sherlock, "A Turbulent Spectrum: Mormon Reactions to the Darwinist Legacy," *Journal of Mormon History* 5 (1978), 33–59.

39. See Ronald N. Giere, *Explaining Science: A Cognitive Approach* (Chicago: University of Chicago Press, 1990), 45–46; and Susan Haack, " 'Realism,' " *Synthese* 73 (1987): 275–99.

40. See Leonard Arrington, "The Intellectual Tradition of the Latter-day Saints," *Dialogue* 4, no. 1 (Spring 1969), 13–26; Davis Bitton, "Anti-Intellectualism in Mormon History," *Dialogue* 1, no. 3 (Autumn 1966), 111–34.

41. "Science and Religion," *Millennial Star* 60 (1 December 1898), 761.

42. "Science Proves 'Mormonism,' " *Deseret News* (1899), reprinted in *Millennial Star* 59 (23 March 1899), 182–83.

43. Gary B. Deason, "Reformation Theology and the Mechanistic Conception of Nature," in *God and Nature*, ed. Lindberg and Numbers, 167–91 (170).

44. The literature documenting this view is extensive. It is not, however, unanimously encouraged, let alone accepted as valid theologically; see, for example, David Bohn's "No Higher Ground: Objective History Is an Illusive Chimera," *Sunstone* 8, no. 3 (May/June 1983), 26–32. Both Mormon and non-Mormon critics have occasioned a full-fledged naturalism; see for instance Klaus Hansen, *Mormonism and the American Experience* (Chicago: University of Chicago Press, 1981); Jan Shipps, *Mormonism: The Story of a New Religious Tradition* (Urbana: University of Illinois Press, 1985); and Dan Vogel, *Religious Seekers and the Advent of Mormonism* (Salt Lake City: Signature Press, 1988).

Also see Howard C. Searle, "Early Mormon Historiography: Writing the History of the Mormons, 1830–1858" (Ph.D. dissertation, UCLA, 1979), 440–49.

45. For the late nineteenth century, see Moore, *The Post-Darwinian Controversies*, passim. In the twentieth century the appearance of Protestant neo-orthodoxy in the 1920s and neo-evangelicalism in the 1940s has tempered this division. In essence there are at least six branches of Protestantism today: Pentecostals and fundamentalists on the ideological right, evangelicals, mainline and neo-orthodox Protestants in the middle, and liberal Protestants on the ideological left. See Sydney E. Ahlstrom, *A Religious History of the American People* (New Haven: Yale University Press, 1972), 873–1096; and William G. McLoughlin, *Revivals, Awakenings, and Reform: An Essay on Religion and Social Change in America, 1607–1977* (Chicago: University of Chicago Press, 1978), 141–216.

46. Eiseley, *The Firmament of Time*.

47. Reynolds, "The Conflict between Science and Superstition," 270.

48. Editor, *Improvement Era* 11 (January 1908), 178.

49. Frederick J. Pack, "Natural and Supernatural," ibid., 178–86. Also see Widtsoe, *Joseph Smith as Scientist*, 30–38.

50. Mormon publications that covered scientific topics were *The Evening and the Morning Star*, the *Messenger and Advocate*, the *Times and Seasons*, the *Wasp*, the *Seer*, the *Millennial Star*, the *Contributor*, the *Juvenile Instructor*, the *Improvement Era*, conference reports republished in the *Journal of Discourses*, and numerous pamphlets, broadsides, and eventually books published by and in defense of the church.

# 2

---

# Developments in Modern Science

In a discussion concerning the relationship of science and religion it is paramount that one understand the basic overall thrust of the development of science as a *human* enterprise. Because issues dealing with the truth claims of both science and religion are central to an alleged conflict between these two human institutions, one must appreciate questions such as: (1) Where do influential scientific ideas come from? (2) What gives them their special authority and appeal? (3) To what extent do they remain the same ideas as they become effective in the larger culture? (4) If the influence of ideas is not literal, in what sense is it really due to the science to which it is imputed? (5) How does science, as a collection of ideas, change or grow?[1] By reflecting on these and subsidiary questions, we will be able to assess the relationship of Mormonism to science.

A variety of studies published in recent years has shown that, following Joseph Smith's lead, Parley P. Pratt, his brother Orson, and B. H. Roberts were the most influential Mormon thinkers to shape Mormon theology.[2] In this context it is frequently alleged that Joseph Smith was a mechanist and philosophical materialist, that Parley P. Pratt espoused a philosophical holism, and that Orson Pratt held an eclecticism that combined both views.[3] Given the assumptions of their *scientific* milieu, these claims are nevertheless misleading. Although Mormons never developed a natural theology (theology deriving its knowledge of God from the study of nature rather than from revelation), their view of the cosmos is consistent with the science that emerged beginning in the seventeenth century. Therefore, in order to place Mormonism into this "scientific" context, we will examine, however briefly, the roots of modern science in light of the questions suggested above.

### The Foundations of Modern Science

Although the roots of modern science are found in the rationalism that developed in Greek antiquity as well as in the several cosmologies that emerged in ancient Greece and the Middle Ages, the foundations of modern science—empirical, conceptual, and methodological—derive primarily from the seventeenth century, the period historians call the Scientific Revolution.[4] During this century three worldviews vied for dominance: the organic or Aristotelian, the mystical or Hermetic, and the mechanical.[5]

In the early years of the seventeenth century the dominant view of the world was medieval Aristotelianism. In this scheme the world (or 'universe' in the parlance of modern science) was divided into two mutually exclusive regions: (1) the sublunar region that included the earth and the region of space up to the moon, and (2) the supralunar region beginning with the moon upwards. Motion in the physical world, that which occurs in and about the earth, was further divided into two types—natural and violent—the former either straight up or straight down, the latter lateral or tangential. Moreover, according to the Aristotelian principle that "everything in motion is moved by another," an object in motion requires a causative agent. Thus, the mover was either an external physical object itself or an internal agent, such as a 'final cause' or an inherent quality.

Ontologically, the world was explained in terms of 'primary' and 'secondary' qualities or forms. The primary qualities were the four Aristotelian elements—earth, air, fire, and water—out of which the Earth, as a corruptible and changeable place, was composed. The supralunar region was incorruptible and composed entirely of ether, the quintessential element. The only change allowed in the supralunar region was circular motion, as evidenced by the wandering planets. All the phenomena in the world were constituted from these five elements. Secondary qualities, such as color, shape, odor, motion, and so forth, were those that changed over time and place. The change in these qualities or forms ('intention and remission' in medieval terms) occurred as temperature, season of the year, environment, and so forth changed. For example, motion—something which an object may possess—was a secondary quality in Aristotelianism, and an attribute which would diminish completely as the object came to rest. Whereas a state of motion was always ephemeral, the state of complete rest was ontologically superior.

In order to support this metaphysics, a physical world picture representing the mechanics of physics and astronomy had been developed. Physics explained the workings of physical objects on and near the earth, and astronomy explained the motions of the sun, the moon, planets, and the stellar region generally. All celestial objects rotated in a circular fashion about the earth because, as the center of the world, the earth was the heaviest, and the only corruptible, object. The celestial machinery was driven by transparent, yet solid, crystalline spheres rotating in perfect circles.[6]

Although Aristotelianism was dominant, contramedieval philosophies, principally Neoplatonism and Renaissance Naturalism or Hermeticism, coexisted with late medieval Aristotelianism. Though they had never been entirely suppressed by the medieval church or by Aristotelianism itself, during the fifteenth century a resurgence of occult philosophies occurred. Proponents of the occult viewed reality strictly in qualitative terms, where physical objects and spiritual or magical powers were intimately intertwined. For example, the lodestone, or magnet, was controlled by some hidden powers, the effects of which were demonstrable but not directly perceivable.

Although modern science has important roots in both Aristotelianism and the Hermetic tradition, its rise in the seventeenth century was established primarily on a *mechanical* conception of nature. The first dramatic event upon which the mechanical conception of nature later built was the introduction of the heliocentric cosmology by the Polish philosopher Nicholas Copernicus (1473–1543) in his *De Revolutionibus Orbium Coelestium* (1543). In 1572 and 1577 the Danish astronomer Tycho Brahe discovered a stellar nova and a new comet, respectively, that, combined, demolished the machinery of the Aristotelian crystalline spheres.[7] In addition, during the first decades of the seventeenth century extremely difficult and prolonged labors by the German astronomer Johannes Kepler produced brilliant discoveries of the (empirical) laws of planetary motions. Particularly important were Kepler's observations that planets orbit in ellipses, not in perfect circles as Aristotle had taught, and his 'law of areas' that allowed for precise calculations of planetary motions. Although only a handful of advocates supported the heliocentric hypothesis by century's end, during the first half of the seventeenth century no one in Europe was unaware that a major conceptual challenge—if not a revolution—was afoot.

The year 1564 marks the anniversary of two of the greatest events of Western intellectual thought: the birth of William Shakespeare, who

ushered in a radical transformation of the English language in Eliza-
bethan England, and the birth of Galileo Galilei, whose ideas placed
modern science on its central foundation. A militant supporter of Copernicanism, Galileo laid the foundations of modern physics with his
complete denial of Aristotelianism and his mathematical explication of
the new cosmos, including, above all, his discovery and advocacy of
the principle of inertia. Whereas in Aristotelianism terrestrial motion
is either vertical or tangential (but not both simultaneously) and something which objects *possess*, with Galileo motion is nearly always compounded and something describing only the *state* of an object. Thus,
argued Galileo, the inertia of an object will continue that object in a
straight line unless, or until, an external force alters its trajectory.[8]
Moreover, an object does not possess inertia or velocity as a secondary
quality. In radical contrast to Aristotelianism, in which motion is an
ephemeral attribute from which all terrestrial objects constantly seek
their place of rest, for Galileo motion is simply a state, and whether
an object is actually in motion or not is wholly irrelevant. Indeed, an
object at *rest* is only in a special state of the more general condition
of that object in motion.

During the first several decades of the seventeenth century, Aristotelianism as a scientific philosophy was in retreat; by the end of the
century it was defunct. During the same period, only the most esoteric
philosophers held that Hermeticism was still a viable cosmic view. As
a result, the new scientific discoveries of Copernicus, Kepler, Galileo,
and others created an enormous philosophical vacuum. As those who
were scientifically and philosophically sensitive began to realize, what
was needed was a new conception of nature, one that would provide
a metaphysics demanded by these and other discoveries. As a result,
by the middle years of the seventeenth century there emerged almost
spontaneously several mechanical philosophies, each proposed independently by the philosophers Marin Mersenne, Pierre Gassendi,
Thomas Hobbes, and, above all, René Descartes.[9] These 'mechanical
philosophies' and other associated developments combined to invalidate utterly the Aristotelian world picture.

In its various formulations this mechanical philosophy attempted to
separate the psychic and the spiritual from the physical and the material; they allowed only for the autonomy of material causation. Thus
physical nature and the phenomena we observe consist only of particles
of matter in motion. In a word, nature is only a complex machine. By
excising mind, spirit, and psychic activity from the world generally,
the mechanical philosophy relegated these nonmaterial activities exclusively to God and the human mind. Above all, the mechanical phi-

losophy provided a metaphysics that explained (supported!) the principle of inertia as well as an ever increasing number of other discoveries.

Although modern science, both physical and biological, was utterly transformed by the conceptual revolution of the seventeenth century, the sciences of astronomy, cosmology, and physics particularly were restructured philosophically and placed on the foundation where they have rested ever since. In contrast to much of Greek and medieval science, the new science of the seventeenth century also discovered the experimental method and achieved a level of mathematical sophistication that far surpassed its roots in antiquity.

## Isaac Newton and Modern Science

The seventeenth-century cosmos has come to suggest a rationally ordered system based on knowable natural law in which all events in the universe can be understood in cause-and-effect relationships. In its modern form this worldview is often associated with the ideas of Isaac Newton (1642–1727). Frequently referred to as 'Newtonian' in the eighteenth century, however, the mechanical view of the cosmos does not entirely represent the views either of Newton himself or of most members of the early modern scientific community.

In 1665, while much of England was being ravaged by the plague, Newton, who was then an undergraduate at the University of Cambridge, returned to the family home in Woolsthorpe. While in seclusion here, a fact which would hardly bear notice had that period not proved to be his annus mirabilis, he invented the calculus, discovered the heterogeneity of white light, and formulated a mathematical law of gravity. Although these scientific insights were still virgin and remained to be developed and unfolded to the world throughout subsequent years, the productions of that brief period were to establish Newton's reputation upon the granite foundation it still enjoys.

Although Newton's own scholarly contributions are legion, his studies in mathematics, optics, and astronomy occupied only a small portion of his time. In fact, most of his mental energies were devoted to church history, theology, the chronology of ancient kingdoms, prophecy, and alchemy. Indeed, he infrequently turned to mathematics, he scarcely touched optics, and he ignored celestial mechanics altogether except for an intense investigation for two and a half years in the middle of the 1680s. Following his brief scientific flirtation in Woolsthorpe in 1665 and 1666, Newton launched himself for the next thirty years into alchemical studies that became his most consuming scientific activity.

Today, leading Newton scholars concur that Newton's alchemical studies became an essential ingredient in the rational science that he eventually constructed.[10]

When thinking of Isaac Newton most people think of rational science in which the world is not only controlled but also explained by forces that are scientifically testable. The fact of the matter, however, is quite different. Newton's world, although explicitly and dominantly mechanical, was saturated both in public and in private with an undercurrent of mysticism.[11] Newton himself was imbued with the Paracelsian quest for the "philosopher's stone." To be sure, he was an experimenter, but first in the Hermetic tradition.

Newton did have a strong belief—an important part of his Renaissance heritage and of the seventeenth-century's intellectual furniture—in an original wisdom or knowledge of the ancients that had been subsequently lost. By applying pure thought, Newton believed, he could partially recover the secrets of the ancients. Consequently, he developed lifelong interests in church history, theology, chronology, and biblical prophecy. But the recovery would be incomplete unless the conclusions could be validated with external, objective events. Here, Newton surmised, alchemy could provide the experimental methods needed for verification. Experimental alchemy, for Newton, was the absolutely essential step needed to uncover the secrets of nature and the hidden treasures of the ancients.[12]

Newton's emphasis on alchemy as the hidden door of the philosopher's stone assumes the existence of a mystical worldview. In seventeenth-century terms this world was provided by the Hermetic tradition. The Hermetic tradition refers primarily to three closely related aspects of a conception of nature shared by every Hermetic philosopher. First, nature was seen as active. Thus individual bodies possess their own sources of activity whereby they set themselves in motion and perform their specific acts. Second, nature was seen as animate. Thus there was no meaningful distinction between the animate and the inanimate; nature was perceived as an organic whole. Consequently, metals grow in the earth, and the experimental alchemist, with the knowledge of the true adept, could accelerate and complete the process in nature and organically transmute base metals into gold. Third, nature was seen as psychic. Thus projecting mind onto nature, bodies exert influences on each other, causing an attraction of like for like. All three aspects of the Hermetic conception of nature—activity, life, and perception—were intertwined with spirit as their locus and carrier. In the Hermetic philosophy, as in all magical traditions, the spiritual and the physical thoroughly penetrated each other and did not stand in sharp distinction.[13]

Until recently, historians of science have been virtually unanimous in seeing this Hermetic philosophy and the mechanical philosophy as antithetical, not only in basic premises but also in the role each played in the emergence of the modern scientific enterprise. The most recent scholarship on the seventeenth century, however, has suggested that mechanical conceptions, disguised by a mechanistic veneer, can only be explained with a Hermetic understanding. In Newton, Hermetic notions shaped his early philosophy of nature and were crucial in the development of his concept of force, which became the central element in modern physics (mechanics and dynamics). This is particularly ironic, because the attractions and repulsions of the Hermetic philosophy of nature are occult virtues, utterly anathema to the mechanical philosophy.[14]

Consider, for example, one of Newton's early speculations on nature, "An Hypothesis Explaining the Properties of Light," sent to the Royal Society in 1675. Representing a sort of alchemical cosmogony, the hypothesis posited the existence of a universal aetherial substance in which the earth, as a vast alembic, perpetually distills its contents into an aetherial spirit which precipitates only to be distilled again. The ether and the aetherial spirits diffused through it were used by Newton as categories to explain a range of active phenomena, including gravity, magnetism, static electricity, the cohesion of bodies, elasticity, heat, sense perception, muscular motion, and vegetable growth.[15]

Within a decade, however, Newton reformulated his aetherial speculations and attempted to excise its overt Hermetic flavor by recasting it into an ether that became explicitly a mechanistic member of his mechanical universe. Although the animism and the active principles of the Hermetic tradition might be disguised, Newton was unable wholly to assimilate it into a strictly mechanical system as the seventeenth century understood it. In the end, the unassimilated Hermetic elements forced Newton into an expanded view of the mechanical universe that (1) reveals its Hermetic heritage by building occult concepts permanently into the structure of modern science and (2) elevated the mechanical philosophy of nature to its mature, modern formulation.

The attractions and repulsions of the aetherial material, as well as active principles generally, Newton ostensibly reduced to his concept of force, in which objects that are physically separated act upon one another from a distance. Content to speak of such forces of attractions and repulsions on the scientific level, however, Newton never believed in the ontological reality of action-at-a-distance. Although the attraction of like for like and the rejection of unlike represented the occult sympathies and antipathies of the Hermetic tradition, attractions and repulsions were only apparent. In Newton's world, which was actually saturated with an animistic flavor, the motions of bodies result from the immediate agency of God, whose infinite presence constitutes infinite space and who, ultimately, controls passive matter. In this way Newton was able to guarantee an active, necessary role for God—to wit, God was the ultimate cause of all action, whether forces between celestial bodies or the antipathies and sympathies evident in this world.[16]

Fashioned in the alchemical world of Hermeticism, Newton's concept of universal gravitation was in an embryonic state in 1679. Not until the early 1680s, when he began in earnest to develop the mathematics of celestial physics, did he begin to realize that Kepler's laws of planetary motion were valid only if every particle of matter attracts every other particle uniformly. That view could not have flouted mechanical sensibilities more openly. When Newton finally developed his system of the world, he transformed the distinctly Hermetic idea of occult attractions into his concept of universal gravitation. Using his immense mathematical prowess, Newton wrote his *Principia Mathematica* (published in 1687), a work critically stimulated by his alchemical speculations, that culminated his mathematical understanding of celestial physics and that eventually revolutionized seventeenth-century thinking, placing science forever on the track that led directly to its modern formulation.

When the work was published, however, it was greeted with charges of occultism. In the spirit of seventeenth-century rationalism, numerous mechanists, who had consciously excised the occult from their science, found Newton's "force" antithetical. Christiaan Huygens, Newton's great nemesis, did not care what ultimate views Newton espoused, "as long as he doesn't serve up conjectures such as attractions."[17] In this sense, the mechanists felt that their disdain for Newton's work was justified. Ironically, however, by adding force to the mechanical universe hitherto populated only by inert matter in motion, Newton enabled science to transcend the level of scholastic rhetoric and to reach the level of quantitative rational mechanics. Newton's alchemical world allowed precisely those factors that guaranteed the eventual success of modern science. In the end, however, the world of the Hermetic philosopher was rejected, and eighteenth-century Enlightenment forgot the occult heritage that became modern science under the guise of rational thinking. Newton's science was not a mere disguise for alchemy. His Hermetic leanings, however, allowed Newton to break radically from the prevailing mechanical philosophy and enabled him to develop his science to the highest levels of achievement.[18]

## Newton in the Age of Reason

A logically consistent understanding of nature had been developed by Aristotle during Greek antiquity and later by his medieval intellectual followers. The emergence of Hermeticism and the demonstration during the sixteenth century that the Aristotelian conception of the cosmos was no longer tenable created a need for both a new physics and a new astronomy to account for the workings of the world. Although usually referred to as Newtonian, the physics was constructed by a host of brilliant seventeenth-century natural philosophers, including such giants as Galileo, Descartes, Huygens, and, of course, Newton himself.[19] Astronomy was the creation of equally perceptive thinkers, particularly Copernicus, Tycho, Kepler, as well as Galileo and Newton.[20]

Modern science, as we know it today, began with the reformulation of the cosmos during the seventeenth century. To support this new vision of reality, a new metaphysics was created that opposed the subjective and the qualitative views of the physical world. In the new philosophy of the seventeenth century only the empirically observable or, at least, the empirically conceivable were admittable as scientific reality. As a result, over the course of the seventeenth century a mechanical philosophy of nature, in which the only ontological realities were matter in motion, emerged as the dominant interpretation of the world. Explanations were thus admitted only if they could be formulated in terms of brute, inert matter in motion. No other principles of scientific causation were allowable. Aside from Newton and a few other scientific thinkers, Galileo, Descartes, Huygens, and most others fundamentally opposed the Hermetic (qualitative, magical) view of nature. Ironically, as we have seen, these two ontological realities (inert matter and motion) were insufficient to complete the mechanical philosophy of nature. In order to explain the machinery of both the heavens and the earth, Newton introduced a third ontological reality into his world system—universal gravitation. Although Newton eventually rationalized his world system fully, he did so by bringing into the very heart of the mechanical philosophy the central feature of the occult.

By the end of the eighteenth century Newtonianism was fully accepted by nearly all European and American scientists. Indeed, Newton became the dominating motif—both as symbol and method—defining what it meant for the eighteenth century to become the Age of Reason. Whether English, French, German, or American, virtually all Western thinkers sought enlightenment as they attempted to emulate Newton. Thus, the greatest expositor of Newton's vision of the cosmos was the French physicist Pierre Simon de Laplace; the most important philosopher of the Newtonian world was the German idealist Immanuel Kant; and the most perceptive theoretician of electrical phenomena was the American Benjamin Franklin.

Although universal gravitation was the central concept of Newton's physics, the effect of gravity was only one of a number of phenomena in which forces were clearly in control. Other "forces" that needed explanation included electricity, magnetism, heat, light, and chemical affinities. Yet Newton had adopted two radically different approaches for the explanation of natural phenomena. On the one hand, mostly in the first edition of his *Principia* and in "Queries," 25–31 of the second edition of his *Opticks* (1706), he espoused an approach that explained the operations of nature in terms of a particulate structure of matter within a framework of force acting-at-a-distance. Using the *Principia* as his methodological guide, Newton emphasized very abstract concepts understood *mathematically;* he provided no *physical* mechanism to account for this action-at-a-distance. On the other hand, in the "General Scholium" of the second edition of the *Principia* (1713) and in his third edition of the *Opticks* (1717), Newton introduced a nonmathematical, ethereal medium in order to explain this force acting-at-a-distance.

In his first conceptual model, space is mostly empty and forces act across this space, whereas in his second model, space is filled with an ether which through direct action accounts for natural phenomena. Newton's ether, however, is not the Cartesian plenum that acts by contiguity and transference of motion. Rather, his ether is comprised of particles repelling each other. Unfortunately, Newton gave no mathematical demonstration of how the elastic forces between the particles could produce any of the observed attractions and repulsions needed to account for these additional phenomena. Also, because Newton's ether is composed of particles repelling each other, it is the embodiment of the action-at-a-distance problem which it pretended to explain. Even so, it did not constitute a compromise with the basic premise on which Newton's philosophy of nature had rested for the more than thirty-five years since the publication of the *Principia*. The net result was that Newton, who came to personify the correct approach to the study of nature, provided two conflicting alternatives for eighteenth-century natural philosophers looking for an authoritative (Newtonian) theory of matter to account for the phenomena of nature. Consequently, two competing scientific traditions, both sanctioned by Newton's reputation, emerged in the eighteenth century.

These two traditions, known in contemporary parlance respectively as 'mechanism' and 'materialism', competed for scientific dominance during the next century.[21] The materialists, however, could not begin their research program until after the publication of the 1717 edition of the *Opticks*, but by then the mechanists were well established. The mechanist tradition was mostly characterized by a severe mathematical approach which, after about the middle of the eighteenth century, produced few positive results. Partially as a result of this general lack of success, sentiment among physical scientists shifted toward the materialist tradition. Eschewing a mechanistic approach, the materialists adopted the view that different 'fluids' accounted for all physical phenomena. Because only fluids required investigation, experimentation, rather than recourse to abstract mathematical ideas, became their method of choice.[22]

By the middle years of the eighteenth century mechanistic or materialistic explanations had been suggested for all natural forces. In the course of these deliberations science in both traditions became a coherent enterprise. Despite the fact that Newton equivocated in his ultimate metaphysics, whatever the ontological basis of 'force', the idea of a universal gravitation was justified in both the mechanistic and the materialistic traditions. In short, science in the eighteenth century became thoroughly Newtonian.

## The Scientific Revolution of the Nineteenth Century

Each of the fluids proposed by the materialists corresponded to a different physical phenomenon—electricity, magnetism, heat, light, chemical affinity, and gravity—and each required a different ontological basis. Although adopting Newton's mathematical approach to science, the mechanists assumed a Cartesian metaphysics yet refrained from overt speculation on the ultimate nature of reality. By and large, they continued to emphasize a mathematical methodology and positivist science. Materialists, who always felt uncomfortable with things mathematical, unabashedly accepted a Newtonian metaphysics but engaged science experimentally. Both mechanists and eventually materialists, however, found this multiplication of realities utterly anathema. Moreover, both groups also found their inability to provide ultimate explanations and equally to be guided by a sufficiently rich and placid metaphysics a crucial hindrance in their understanding of nature.

While both materialists and mechanists consciously eschewed discussion of nature ultimately in terms of force, beginning with Gottfried Wilhelm Leibniz in the seventeenth century and the philosophers Roger Boscovich and Immanuel Kant in the eighteenth century, scholars grew increasingly aware of the primacy of force as the basic conceptual category for organizing natural phenomena. They argued that all natural phenomena—whether, for example, planetary motion, electrical activity, or vital biological principles—are manifestations of the *same* underlying force; therefore, this force must ultimately be unified and whole. Consequently, by the early years of the nineteenth century there had emerged three sets of explanatory notions of natural phenomena—all of which were mutually contradictory both scientifically and philosophically: (1) imponderable fluids, (2) positivist, mathematical science, and (3) the primacy and unity of forces. The resolution of these conflicting worldviews led to the development of a new metaphysics, radically different from both Newtonian and Cartesian views, that came to dominate nineteenth-century science. The development of this new worldview, called 'field theory', constitutes partially what historians call the Second Scientific Revolution.[23]

Field theory, like Newtonian and Cartesian metaphysics, attempted to solve the mystery of the universe and to discover the secrets of nature. Beginning in the 1820s, field theory sought answers to a set of fundamental, age-old problems, such as: (1) How does one body act on another? (2) Why does one body push another along instead of penetrating it? (3) How is a magnet able to cause a piece of iron some distance away to move, and, by extension to celestial phenomena, (4) how does gravitation actually cause planetary bodies to move in their regular orbits? Though using mostly nonmathematical approaches to the study of nature, during the first half of the nineteenth century those who adopted either explicitly or implicitly a field-theoretic approach achieved impressive scientific results, as in the work of Thomas Young (wave theory of light), Hans Oersted (electric-magnetic connections), and particularly Michael Faraday (electromagnetic induction).[24] Using the field-theory idea so forcefully advanced by Faraday, during the 1860s and '70s the Scottish scientist James Clerke Maxwell brilliantly proposed mathematical unification theories that fully interrelated electricity, magnetism, and light.

Whereas Newtonianism required that force must be localized within a material body, in radical contrast to both Newtonianism and Cartesianism, field theory posited the existence of nonmaterial fields of force that exist in space even where there is no matter and through and by which natural phenomena were thought to be manifested. Moreover, in contrast to Newtonian action-at-a-distance theories, in which action is instantaneous, field theories predicted that all actions of one body on another take time. The crucial experiment that decided between these conflicting interpretations of nature—that is, whether certain actions move instantaneously or not—was designed and performed by the German scientist Heinrich Hertz in 1887. In that experiment field theory found its greatest triumph with the discovery of electromagnetic waves (radio waves).[25]

Developed toward the end of the eighteenth century, the unity-of-forces idea, which eventually led to field theory, posits that force may be manifested in numerous ways, such as in electricity, magnetism, light, heat, and mechanical work. Consequently, from the 1820s to the mid-1840s at least a dozen scientists considered the possibility of the convertability of these various forces. As a result, the concept of unity of forces also laid the groundwork of the principle of the conservation of energy (forces), codiscovered independently during the 1840s by the Germans Robert Julius Mayer and Hermann von Helmholtz.[26] Although Mayer's discussion of the nascent principle was highly qualitative, Helmholtz placed the conservation of energy idea on a firm mathematical foundation. Consequently, by the end of the nineteenth century most of the hitherto qualitative material forces of nature except gravity had been reduced to a single manifestation—using, however, mostly abstruse mathematical ideas.

In the two centuries following Newton many important achievements in the physical sciences dealt frequently with the development of a strictly mathematical approach to problem solving and theory formation. This was particularly crucial in the influential French school of physics that developed the theoretical foundations of resisting media (fluids), celestial mechanics (planetary theory), cosmology (nebular hypothesis), electricity and magnetism (Ampère), and the wave theory of light (Fresnel). As impressive as these results were, however, the unity-of-forces and field-theory ideas eventually led science to its highest levels of understanding during the nineteenth century. Using the new metaphysics, nineteenth-century physical science produced two great inventions—electromagnetism and thermodynamics (which includes the kinetic-molecular theory of matter). The key research problems entailed in these two programs were (1) the propagation of light and heat radiation and their interaction with matter, (2) the connection between electricity and magnetism, and (3) the balance and flow of energy.[27] (Problems dealing with gravity did not receive their current resolution until after the work of Albert Einstein in the early decades of the twentieth century.)

Although mathematics became the lingua franca of post-Newtonian science and a radically new metaphysics came to dominate, the approach to science during these two centuries was still Newtonian: nature was rationally understandable and open to repeated inquiry, explainable, and predictable. All events in nature had a cause and produced an effect. In short, nature was transparent, and truth and understanding were obtainable.

## Copernicanism and the Plurality of Worlds

The pre-Socratic atomists of ancient Greece, Leucippus and Democritus, first explored the idea of a plurality of worlds. As initially conceived, this tradition espoused a plurality of *kosmoi*—cosmic systems composed of an earth, planets, and fixed stars. It was the multiplication of world as *kosmos* that preoccupied these early Greeks; the modern notion of extraterrestrial life was of minor consideration.[28] In time the view developed that circling the myriad of stars in the heavens are worlds without number—planets, inhabited by rational, sentient beings, not unlike the earth with its human population.

With the rise of Aristotle's system in the fourth century B.C., the plurality concept became largely dormant until the rediscovery of the Greek atomists in the middle years of the European Renaissance. The image of an earth-centered universe, supported during the Middle Ages by religion, philosophy, and Aristotelian science, suggested a unique position for the earth in the cosmos. Thus it was conceptually difficult to argue in favor of coexisting multiple world systems. In fact, if the earth were the singular center of the universe, it seemed absurd to suggest that the stars in the firmament are suns comparable to our own, let alone that they possess inhabited planets. After all, went the argument, life in the universe is found logically and empirically only at the center. The expanse of the cosmos and its perimeter is reserved for God, angels, and quintessential substances.[29] Not until the middle years of the sixteenth century, after Copernicus presented his astronomically tenable heliocentric cosmology, did pluralistic ideas first begin to emerge in the West.

Heliocentrism is not necessarily essential to the plurality-of-worlds debate, but placing the sun instead of the earth at the center of the universe made it possible for the plurality of worlds idea to emerge.[30] During this period the revival of early Greek Epicurean cosmogony in the new science of the seventeenth century suggested that, because atoms are the ultimate agents of causality and infinite causes must have infinite effects, the formation of an infinite number of worlds was demanded by the fortuitous coalescence of an infinite number of atoms. The invention of this (mechanical) philosophy of nature provided the metaphysical support required by radical Copernican cosmology. Together, Copernicanism and atomism altered the intellectual climate of Western thought in new and creative ways. Although several generations passed before people were able to understand fully the theological and scientific nature of this emerging worldview, with the subsequent refinement of heliocentrism by Kepler, Galileo, Descartes, Newton, Huygens, Leibniz, and a host of other seventeenth-century natural philosophers, the concept of astronomical pluralism began to develop in earnest.

Belief in the validity of the plurality of inhabited worlds eventually filtered down to the popular level with numerous editions and translations of Bernard de Fontenelle's widely known 1686 treatise *Conversations on the Plurality of Worlds*, the first successful treatment of pluralism intended for general dissemination.[31] During the Enlightenment belief in this doctrine became pervasive. Although some thought its reality improbable, many accepted it. The European natural philosophers Thomas Wright, Immanuel Kant, Johann Lambert, and later William Herschel, the most important observational astronomer of the late eighteenth and early nineteenth centuries, all advocated astronomical pluralism.[32] In eighteenth-century America the doctrine was advanced by a number of prominent figures. At Harvard, for instance, teachers explained that all planets of the solar system, not just the earth, are inhabited; otherwise, they argued, these planets would have been created in vain, which God, of course, would not do. Within a decade the pluralist view entered the curriculum and remained an essential ingredient of the theological training of Yale ministers. Thanks especially to the Yale theologian Timothy Dwight, several generations of ministerial students were fed the pluralist diet. Such literary and religious figures as Ezra Stiles and the American Samuel Johnson spoke of the morals "those inhabitants of this earth and the planetary starry universe" display, while both David Rittenhouse, the colonies' foremost astronomer, and Benjamin Franklin, American's foremost scientist of the eighteenth century, espoused the doctrine.[33] By the end of the eighteenth century, pluralism was advocated by leading scientists and natural philosophers, who wedded the doctrine to the current scientific theories of the day. For instance, by the turn of the nineteenth century Pierre Simon de Laplace, the most gifted mathematical astronomer since Newton, argued that multiple world systems had to exist because the formation of stars and planets, resulting from the rotation, contraction, and condensation of the primeval solar material and gases, was a natural process throughout the universe.

## Naturalism and Post-Newtonian Cosmology

The development of the science of cosmology in Western culture can be divided into two periods, before and after the Scientific Revolution of the seventeenth century. Roughly before 1600, as a science cosmology was primarily *planetary* cosmology, while after 1800 it increasingly became *stellar* cosmology. There are two basic reasons for this shift. Before the publication of Copernicus's *De Revolutionibus* in 1543, scientists, philosophers, and theologians held the view that the physical universe is composed of the planets of a geocentric solar system surrounded by a sphere of fixed stars. In the century following the publication of *De Revolutionibus,* the machinery of the heavens that had kept medieval (planetary) cosmology intact had been totally dismantled. In the process the stars were eventually raised to the status of physical objects. Galileo's use of the recently invented telescope in 1609 gave astronomers the primary tool needed to study the stars and, for that matter, the whole panorama of heavenly phenomena, including the issue of the plurality of worlds as the stars were increasingly seen as objects similar to our sun.[34]

Following the demise of medieval conceptions of the heavenly realm, scientists from Newton onwards increasingly viewed the stellar regions in naturalistic terms. Using primarily the telescope and the nascent Newtonian theory, late seventeenth- and eighteenth-century astronomers began asking questions concerning the evolution and physical nature of the stars. Although Newton imagined that the universe periodically wound down, thus requiring the active agency of God to reinvest it with the needed force, he further suggested that stars were replenished by falling comets.[35] But it was Newton's younger contemporary, the cleric William Whiston, who most forcefully explored this natural mechanism for stellar development. Whiston went on to argue that the earth itself had been formed from a nebulous comet and that both the Noachian flood and the millennial conflagration have their explanation in the near approach of a comet. Extending this explanation as the principal mechanism for the formation of the earth, the French naturalist Georges Buffon in 1750 suggested that millions of years ago a comet had swept past the sun, tearing material that subsequently coalesced into our planetary system. By the middle of the eighteenth century the European natural philosophers Thomas Wright, Immanuel Kant, and Johann Lambert all independently imagined similar cosmologies in which the stars were spread out in a plane of almost boundless dimensions. By 1755 Kant had suggested that the solar system itself resulted from the rotation and contraction of some sort of primeval stellar material. Although highly innovative and qualitative in nature, these cosmologies addressed crucial philosophical issues.

In 1791 the English astronomer William Herschel, famous for his discovery of the planet Uranus in 1781, first observed nebulous clusterings in the stellar region which he concluded were protostars in the making. Using Herschel's idea and developments of a century of brilliant advances in (Newtonian) celestial mechanics, the gifted mathematical astronomer Laplace proposed the 'nebular hypothesis'—that the formation of stars and planets resulted from the rotation, contraction, and condensation of primeval solar material and gases.[36]

By the end of the eighteenth century advances in optics, telescope making, and Newtonian theory all urged a rigorous and quantitative direction to these concerns. By the turn of the nineteenth century, a variety of post-Copernican stellar cosmologies had been proposed, principally by Descartes, Newton, Wright, Kant, and Lambert. It was William Herschel, however, who first rigorously quantified the observational study of the cosmos.[37] Working under the assumptions that the stars are both equally bright and evenly distributed in space, beginning in 1785 Herschel developed the first observational and quantitative cosmology of the stellar region. Known among astronomers as the 'construction of the heavens'—a program to count and catalogue the positions, brightnesses, and eventually the motions of stars, Herschel's general approach was adopted by all the important astronomers of the nineteenth and early twentieth centuries.

Among those who contributed most to speculative, dynamic cosmologies were Maxwell Hall, Hugo Gyldén, and Edward Schönfeld, while Wilhelm Struve, Herschel's son John, Friederich Argelander, and Truman Safford all added considerably to the discussion of Herschel's research program. As a result, throughout the nineteenth century Herschel's mathematics were enormously expanded and his assumptions fully explored. Although the gains in the technical nature of astronomy were greatly developed, successful scientific cosmologies remained premature. By the first decades of the twentieth century, however, Herschel's modest beginnings culminated in very sophisticated statistical cosmologies developed primarily by the German astronomer Hugo von Seeliger and the Dutchman J. C. Kapteyn. Their cosmological models implied that the Milky Way Galaxy is rather flat, like a saucer, organized around our very own sun with the preponderance of stars near the solar center.[38]

## Themes in Twentieth-Century Science and Cosmology

The statistical models of Seeliger and Kapteyn that had dominated a whole generation of astronomers were eventually overthrown and reformulated during the 1920s by a radically different vision of the observable universe, created principally by American astronomers.[39] In 1918 Harlow Shapley concluded that the Milky Way Galaxy is approximately 300,000 light-years in diameter, with the solar system asymmetrically centered roughly 90,000 light-years from the galactic center.[40] In 1924 Edwin Hubble provided incontrovertible evidence that our Milky Way Galaxy is only one of countless 'island galaxies' in the universe.[41] In 1929 Hubble further demonstrated that all the galaxies within the known universe were receding from one another, suggesting that at one point in the early history of the universe all this galactic mass was indeed localized.

Since then modern theories of cosmogony assert that when the universe began—some fifteen billion years ago—all of the matter contained in the present-day universe was compressed into an incredibly small and enormously dense mass. In order to understand this idea, cosmologists have had to reduce this primordial mass to a mathematical point while ignoring implications of its physical dimensions. But still it is a "something"; therefore, this view does not compel *ex nihilo* assumptions. At some point in time—which marks the beginning of the physical universe—this mass experienced an incredibly rapid expansion, which cosmologists call the big bang.[42] Empirical verification for this view, if true, should be found in the residual radiational remains of this singular event and the subsequent expansion of the original mass. As it turns out, in 1963 Arno Penzias and Robert Wilson discovered the background radiation providing the powerful residual evidence from this genesis event.[43] There is evidence that the present expansion of the universe, now some fifteen billion years in diameter, is slowing down. If this evidence is valid and the deceleration increasingly reverses its expansion to contraction, the universe which began with a big bang will end with a big crunch.

After Albert Einstein introduced his 'special theory of relativity' in 1905, establishing an equivalence between matter and energy, scientists no longer could speak of absolute measures of time, distance, and mass; rather, the events of moving objects—now considered as time dilation, length contraction, and mass compression—could only be measured

Medieval cosmology. Although some of the details varied, the framework of the medieval worldview was always the same: the central earth as the point from which all motion was defined according to Aristotle's physics. This diagram from Andreas Cellarius, *Harmonia Macrocosmia* (1661), depicts the Ptolemaic Aristotle's four elements, with the earth and its water surrounded by the spheres of air and fire, which, in turn, are surrounded by spheres of the seven traditional "planets": Moon, Mercury, Venus, the Sun, Mars, Jupiter, and Saturn. This is followed by the sphere of fixed stars, the ninth celestial sphere, the sphere of the prime mover, and finally the "empyreal heaven, abode of god and of all the elect."

Medieval cosmology. This diagram shows the Ptolemaic system of the world depicting Aristotle's four elements—earth, water, air, and fire—at the center with the seven concentric planets depicted as chariots.

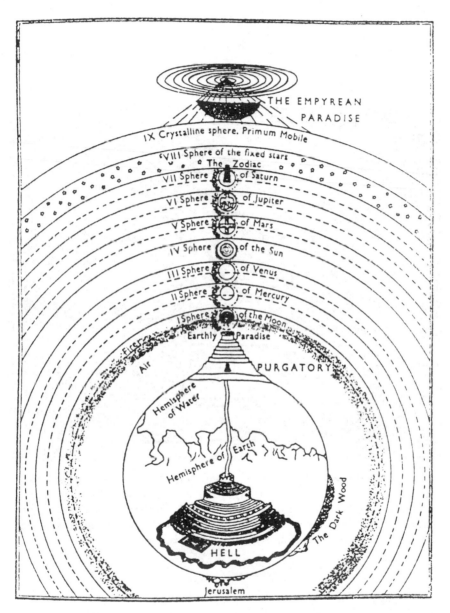

Dante's Scheme of the Universe. A slightly modified form of the medieval worldview as presented in *The Divine Comedy* at the time of Dante (1265–1321)

Medieval worldview. This nineteenth-century rendition (1888) by the French astronomer Camille Flammarion urges the view that what lies beyond the sphere of stars is a wondrous universe filled with celestial machinery and other marvels. In the upper left one sees several other worlds, reflecting the growing interest with post-Copernican plurality of worlds.

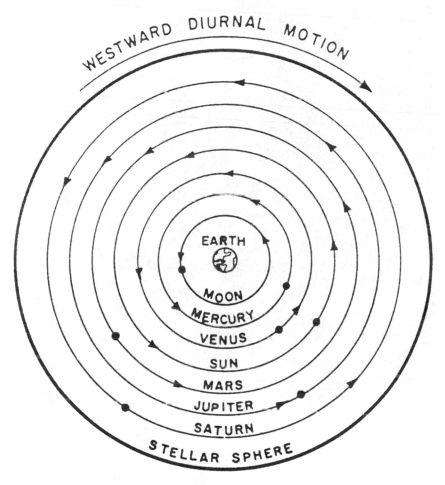

Medieval cosmology. Earth at the center of the world, surrounded by the planets in order of their orbital times

Nicholaus Copernicus (1473–1543), whose cosmological speculations eventually led to the complete denial of two thousand years of Aristotelian science and philosophy. Courtesy of Yerkes Observatory

# NICOLAI CO

## PERNICI TORINENSIS

### DE REVOLUTIONIBVS ORBI
### um coelestium, Libri VI.

Habes in hoc opere iam recens nato, & ædito,
studiose lector, Motus stellarum, tam fixarum,
quàm erraticarum, cum ex ueteribus, tum etiam
ex recentibus obseruationibus restitutos: & nouis insuper ac admirabilibus hypothesibus ornatos. Habes etiam Tabulas expeditissimas , ex
quibus eosdem ad quoduis tempus quàm facilli
me calculare poteris. Igitur eme, lege, fruere.

ἀγεωμέτρητος μδεὶς εἰσίτω.

BIBL.
VNIVERS.
LIPS.

Norimbergæ apud Ioh. Petreium,
Anno   M.   D.   XLIII.

Copernicus's *De Revolutionibus* (1543). This copy is located at the Leipzig University Library.

Copernican cosmology. This seventeenth-century post-Copernican diagram, from Andreas Cellarius, *Harmonia Macrocosmia* (1661), emphasizes the importance of the sun-centered system.

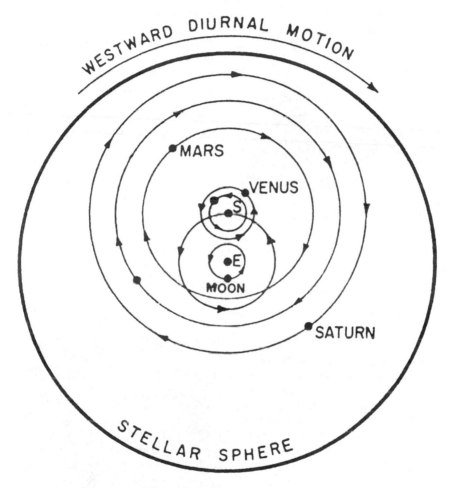

The Tychonic system. The earth is central to the universe, with the moon and sun moving in their old Ptolemaic orbits; the remaining planets rotate about the sun as a common center. Thus Tycho combined features of the Aristotelian-Ptolemaic system with the Copernican.

Johannes Kepler (1571–1630). Kepler combined the speculations of Copernicus with the observations of Tycho to discover the laws governing the orbits of the planets. Thus he provided a vital transition between Tycho's cosmology and that of Newton. Courtesy of Yerkes Observatory

Galileo Galilei (1564–1642). By rejecting Aristotle and adopting the new Copernican cosmology, Galileo was led to discover the physics of inertia which later provided a crucial link to the physics of Newton. Courtesy of Yerkes Observatory

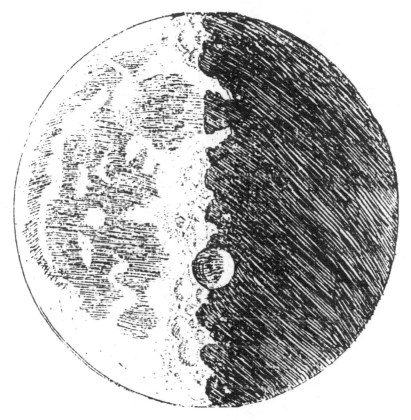

Galileo's first drawings of the lunar surface (1610). Kepler was later to use the circular features on these charts to argue for the existence of lunar inhabitants.

# SIDEREVS
## NVNCIVS
MAGNA, LONGEQVE ADMIRABILIA
Spectacula pandens, fuspiciendaque proponens
vnicuique, præsertim verò

*PHILOSOPHIS*, *atq. ASTRONOMIS*, *quæ à*

# GALILEO GALILEO
### PATRITIO FLORENTINO
Patauini Gymnasij Publico Mathematico
### PERSPICILLI
*Nuper à se reperti beneficio sunt observata in LVNÆ FACIE, FIXIS IN-*
*NVMERIS, LACTEO CIRCVLO, STELLIS NEBVLOSIS,*
*Apprime verò in*
### QVATVOR PLANETIS
Circa IOVIS Stellam disparibus interuallis, atque periodis, celeri-
tate mirabili circumuolutis; quos, nemini in hanc vsque
diem cognitos, nouissimè Author depræ-
hendit primùs; atque

# MEDICEA SIDERA
### NVNCVPANDOS DECREVIT.

VENETIIS, Apud Thomam Baglionum. M DC X.
*Superiorum Permissu, & Privilegio.*

Galileo's *Siderius Nuncius* (1610). In this popular little book Galileo provided
the first detailed observational evidence for lunar craters, the moons about
Jupiter, and the myriad of stars in the Milky Way.

Galileo presenting his telescope to the muses and pointing out a heliocentric system

Frontispiece to the original edition of Galileo's *Dialogue*. While not the three characters in Galileo's dialogue, three figures (from left to right) represent Aristotle, Ptolemy, and Copernicus in dialogue about whose cosmology is the true system of the world.

# DIALOGO
## DI
## GALILEO GALILEI LINCEO
### MATEMATICO SOPRAORDINARIO
#### DELLO STVDIO DI PISA.

*E Filofofo, e Matematico primario del*

#### SERENISSIMO

## GR.DVCA DI TOSCANA.

Doue ne i congreffi di quattro giornate fi difcorre
fopra i due

### MASSIMI SISTEMI DEL MONDO
#### TOLEMAICO, E COPERNICANO;

*Proponendo indeterminatamente le ragioni Filofofiche, e Naturali
tanto per l'vna, quanto per l'altra parte.*

CON PRI      VILEGI.

IN FIORENZA, Per Gio:Batifta Landini MDCXXXII.

*CON LICENZA DE' SVPERIORI.*

Galileo's *Dialogues Concerning the Two Chief World Systems* (1632). The two
systems are the Ptolemaic and the Copernican; he makes brief mention of the
Tychonic system and its variants.

Isaac Newton (1642–1727) established the foundations of modern physics and astronomy. His influence not only affected all areas of science; Newtonianism became synonymous with the proper method for understanding both natural and human phenomena. Courtesy of Yerkes Observatory

# PHILOSOPHIÆ

## NATURALIS

# PRINCIPIA

## MATHEMATICA.

Autore *J S. NEWTON*, *Trin. Coll. Cantab. Soc.* Matheseos
Professore *Lucasiano*, & Societatis Regalis Sodali.

## IMPRIMATUR·

S. P E P Y S, *Reg. Soc.* P R Æ S E S.

*Julii* 5. 1686.

## L O N D I N I,

Jussu *Societatis Regiæ* ac Typis *Josephi Streater.* Prostat apud
plures Bibliopolas. *Anno* MDCLXXXVII.

Newton's *Principia Mathematica* (1687). This enormously influential book provided a synthesis of the physics of terrestrial and celestial phenomena.

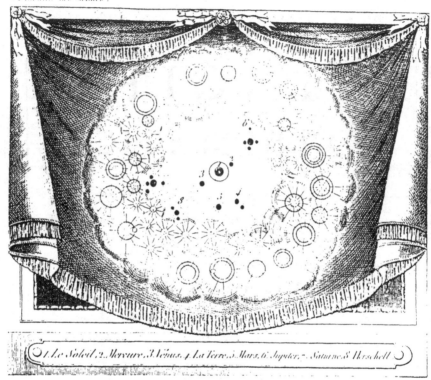

This frontispiece to Bernard de Fontenelle's *Conversations on the Plurality of Worlds* (1686) shows Fontenelle's belief in the plurality of worlds around the sun and the fixed stars. This 1809 American edition depicts the planet Uranus (discovered by William Herschel in 1781).

relative to one another in inertial frames of reference limited by the velocity of light. With Einstein's 1916 'general theory of relativity,' the flat geometry of Euclid became wholly inadequate to explain the macroproperties of the entire universe. General relativity identified fields of force with space curvature, where the presence of matter curves space and bends it round into a hypersphere. In this new universe, shortest distances (and hence "space") are curved by the presence of massive objects, such as stars, where the curvature of space itself may be characterized by the trajectory of stellar light.[44] "Space" itself was no longer described by the mathematics of Euclidean geometry, where the shortest distance between two points is a Euclidean straight line. As a result, space relinquished its three-dimensional Euclidean structure, and "time" relinquished its historical claim as merely a property of temporal reality. Cosmologists now no longer talk of an infinitely extended universe; rather, they talk of a "space-time" continuum in which the spatiotemporal extent of the universe is unbounded but finite and in which the universe is represented, like the *surface* of a Euclidean sphere, with three, not just two, dimensions.

But Einstein's relativity was only a harbinger of the emergence of a new worldview. On the one hand, relativity is a kind of field theory, so when evidence supported Einstein's relativistic theory of gravitation over Newton's, it was a victory for field theory. On the other hand, by the 1890s field theory had become increasingly unable to explain some crucial phenomena, principally discoveries related to 'black-body' radiation—what is called "the scandal of the ultraviolet."[45] To explain this problem, in 1900 the German physicist Max Planck introduced a completely ad hoc notion—the quantum idea. But its break with ordinary physical ideas did not occur until 1905, when Einstein provided the first suggestion of a new worldview incorporating the quantum concept. During the 1920s and '30s Niels Bohr, Werner Heisenberg, Max Born, Louis de Broglie, Erwin Schrödinger, and others developed quantum mechanics into its modern formulation.[46] As a result, field theory was eventually undermined by quantum mechanics, which denies that matter or energy is continuously distributed throughout space, as assumed in field theory.

Even more than the creation of relativity theory, the invention of quantum physics resulted in the total denial of physical determinism and in its replacement by a probabilistic description of the world, in which nature itself, as a physical reality independent of humans as observers, could no longer be guaranteed. In the most widely accepted understanding of quantum mechanics, the so-called Copenhagen interpretation, Danish physicist Niels Bohr argued that the observer and the world were so inextricably connected that "an independent reality in the ordinary physical sense can neither be ascribed to the phenomena nor to the agencies of observations."[47] This means that many of the physical properties of atomic physics did not even exist before the actual act of observation; in other words, the act of observation was necessary to bring these properties into existence! As a consequence, quantum theory and relativity have undermined the classical (pre-twentieth century) dichotomy between the observer and the thing being observed.[48]

Beginning with Newton and Whiston in the seventeenth century through the early years of the nineteenth century, astronomers, though rarely denying the ultimate divine nature of the cosmos, provided a variety of natural mechanisms to account for the evolution and structure of the universe. Moreover, principles of a rationally ordered universe have remained a sine qua non of modern science since its founding in the science of Galileo, Descartes, and Newton. This pertains also to such methodological and philosophical assumptions as universality (uniformitarianism and actualism)—the concept that the universe operates temporally in the same fashion and the same degree, respectively, yesterday, today, and tomorrow, as well as spatially and isotropically in all directions and places. Science today also assumes that nature operates in the simplest way without extraphysical influences, thus assuring a completely naturalistic ethos.

The nineteenth-century belief in cause-and-effect relationships, however, has in this century been replaced with a less transparent and less optimistic view of the universe. Beginning with the development of the atomic theory and the kinetic theory of gases during the latter half of the nineteenth century, science changed radically from its earlier deterministic Newtonian foundation.[49] Although many events can still be understood in this way, not only have twentieth-century developments in astrophysics fundamentally changed our whole conception of the nature of the observable universe, but relativity theory and quantum mechanics have deeply shaped our understanding of the nature of space and of the cosmos generally and of how scientists understand reality altogether.[50]

## NOTES

1. Thomas S. Kuhn, "The Relations between History and History of Science," *Daedalus* (1971), 276.

2. See Leonard Arrington, "The Intellectual Tradition of the Latter-day Saints," *Dialogue* 4, no. 1 (Spring 1969), 13–26; Peter Crawley, "Parley P. Pratt:

The Father of Mormon Pamphleteering," *Dialogue* 15, no. 3 (Autumn 1982), 13–26; David Whittaker, "Orson Pratt: Prolific Pamphleteer," *Dialogue* 15, no. 3 (Autumn 1982), 27–41; and E. Robert Paul, "Early Mormon Intellectuals: Parley P. and Orson Pratt," *Dialogue* 15, no. 3 (Autumn 1982), 42–49.

3. For example, see Whittaker, "Orson Pratt," 32.

4. For two of numerous excellent sources to consult on this topic, see the classic studies by A. Rupert Hall, *The Scientific Revolution, 1500–1800: The Formation of the Modern Scientific Attitude*, 2d ed. (Beacon Press, 1966), and Herbert Butterfield, *The Origins of Modern Science, 1300–1800*, rev. ed. (New York: Free Press, 1965). A superb, but revisionist, interpretation is given by David C. Lindberg and Robert S. Westman, eds., *Reappraisals of the Scientific Revolution* (New York: Cambridge University Press, 1990).

5. The literature on this topic is enormous; see, for instance, Hugh Kearney, *Science and Change, 1500–1700* (New York: McGraw-Hill, 1971).

6. For a superb introduction to medieval science, see David C. Lindberg, ed., *Science in the Middle Ages* (Chicago: University of Chicago Press, 1978); and Edward Grant, *Physical Science in the Middle Ages* (New York: Cambridge University Press, 1978).

7. Victor E. Thoren, *Tycho Brahe, The Lord of Uraniborg* (New York: Cambridge University Press, 1991).

8. Galileo Galilei, *Dialogue Concerning the Two Chief World Systems—Ptolemaic and Copernican*, trans. S. Drake (Berkeley: University of California Press, 1967), 28–50. Strictly, Galileo's idea of inertia is more accurately a "circular" inertia, later corrected by Descartes to "straight-line" inertia. For a highly insightful discussion of these developments, see Stillman Drake, *Galileo at Work: His Scientific Biography* (Chicago: University of Chicago Press, 1978), 25–27, 126–32, and Drake's most recent study, *Galileo: Pioneer Scientist* (Toronto: University of Toronto Press, 1990), passim.

9. See, for example, Peter Dear, *Mersenne and the Learning of the Schools* (Ithaca: Cornell University Press, 1988), and Lynn Sumida Joy, *Gassendi and the Atomists: Advocate of History in an Age of Science* (New York: Cambridge University Press, 1987).

10. In addition to the sources cited below, see Keith Hutchison, "What Happened to Occult Qualities in the Scientific Revolution?" *Isis* 73 (June 1982), 233–53, and Betty Jo Teeter Dobbs, "Newton's Alchemy and His Theory of Matter," *Isis* 73 (December 1982), 511–28.

11. For a wonderfully accessible source on the nature of the mystical worldview during this period, see Allen G. Debus, *Man and Nature in the Renaissance* (New York: Cambridge University Press, 1978); for a technical discussion of the same issues, see Allen G. Debus, *The Chemical Philosophy: Paracelsian Science and Medicine in the Sixteenth and Seventeenth Centuries*, 2 vols. (New York: Science History Publications, 1977).

12. Betty Jo Teeter Dobbs, *The Foundations of Newton's Alchemy; or, "The Hunting of the Greene Lyon"* (New York: Cambridge University Press, 1975), 14.

13. Richard S. Westfall, "Newton and the Hermetic Tradition," in *Science, Medicine, and Society in the Renaissance*, ed. Allen G. Debus (New York: Science History Publications, 1972), 184.

14. For a superb review of the state of contemporary Newtoniana, see Richard S. Westfall, "The Changing World of the Newtonian Industry," *Journal of the History of Ideas* 37, no. 1 (1976), 175–84.

15. Ibid., 189–90, and G. E. McGuire, "Transmutation and Immutability: Newton's Doctrine of Physical Qualities," *Ambix* 14 (1967), 84–6.

16. Westfall, "Newton and the Hermetic Tradition," 192, and David Kubrin, "Newton and the Cyclical Cosmos: Providence and the Mechanical Philosophy," *Journal of the History of Ideas* 28 (1967), 325–46.

17. Christiaan Huygens, quoted in Richard S. Westfall, *Never at Rest: A Biography of Isaac Newton* (New York: Cambridge University Press, 1980), 464.

18. Richard S. Westfall, "The Influence of Alchemy on Newton," in *Mapping the Cosmos*, ed. Jane Chance and R. O. Wells, Jr. (Houston: Rice University Press, 1985), 114.

19. I. Bernard Cohen, *The Birth of a New Physics*, rev. ed. (New York: W. W. Norton, 1985).

20. Alexandre Koyre, *From the Closed World to the Infinite Universe* (Baltimore: Johns Hopkins University Press, 1957).

21. See Robert Schofield, *Mechanism and Materialism* (Princeton: Princeton University Press, 1970), and Arnold Thackray, *Atoms and Powers: An Essay on Newtonian Matter-Theory and the Development of Chemistry* (Cambridge, Mass.: Harvard University Press, 1970).

22. The French, forever suspicious of Newton's materialism, remained steadfast mechanists and, with the work of a legion of brilliant mathematical physicists such as Laplace, eventually dominated much of modern science by the middle years of the nineteenth century. See Thomas L. Hankins, *Science and the Enlightenment* (New York: Cambridge University Press, 1985), and P. M. Harman, *Energy, Force and Matter: The Conceptual Development of Nineteenth-Century Physics* (New York: Cambridge University Press, 1982).

23. Stephen G. Brush, *The History of Modern Science: A Guide to the Second Scientific Revolution, 1800–1950* (Ames: Iowa State University Press, 1988).

24. L. Pearce Williams, *Michael Faraday, A Biography* (New York: Basic Books, 1965).

25. William Berkson, *Fields of Force: The Development of a World View from Faraday to Einstein* (London: Routledge and Kegan Paul, 1974), 213–55.

26. See D. S. L. Cardwell, *From Watt to Clausius: The Rise of Thermodynamics in the Early Industrial Age* (Ithaca: Cornell University Press, 1971); Yehuda Elkana, *The Discovery of the Conservation of Energy* (Cambridge, Mass.: Harvard University Press, 1974); and C. Truesdell, *The Tragicomical History of Thermodynamics, 1822–1854* (New York: Springer-Verlag, 1980).

27. See Stephen G. Brush, *The Temperature of History: Phases of Science and Culture in the Nineteenth Century* (New York: B. Franklin, 1978).

28. Steven J. Dick, *Plurality of Worlds: The Origins of the Extraterrestrial Life Debate from Democritus to Kant* (New York: Cambridge University Press, 1982), 1–12.

29. Medieval speculations on the plurality of inhabited worlds, however, were not entirely unknown. See, for example, Nicole Oresme [ca. 1325–82], "The Possibility of a Plurality of Worlds," in *A Source Book in Medieval Science*, ed. Edward Grant (Cambridge, Mass.: Harvard University Press, 1977), 547–54. Oresme was among the most influential medieval philosophers and theologians.

30. See Steven J. Dick, "The Origins of the Extraterrestrial Life Debate and Its Relation to the Scientific Revolution," *Journal of the History of Ideas* 41, no. 1 (1980), 3–27; Dick, *Plurality of Worlds*, esp. 61–105; and Arthur O. Lovejoy, *The Great Chain of Being* (Cambridge, Mass.: Harvard University Press, 1936), 24–98.

31. Bernard le Bovier de Fontenelle, *Conversations on the Plurality of Worlds*, trans. H. A. Hargreaves (Berkeley: University of California Press, 1990). Others attempted to present pluralism to the public before Fontenelle, but without success. See John Wilkins, *Discourse Concerning a New Planet* (1640) and Pierre Borel, *Discours nouveau prouvant la pluralité des mondes* (1657).

32. See Dick, *Plurality of Worlds*, 159–75; and Simon Schaffer, " 'The Great Laboratories of the Universe': William Herschel on Matter Theory and Planetary Life," *Journal for the History of Astronomy* 11 (1980), 81–111.

33. Herbert Leventhal, *In the Shadow of the Enlightenment: Occultism and Renaissance Science in Eighteenth-Century America* (New York: New York University Press, 1976), 242–47.

34. See Albert van Heldon's introduction to Galileo Galilei, *Sidereus Nuncius; or, The Sidereal Messenger*, trans. by A. van Helden (Chicago: University of Chicago Press, 1989), 1–24.

35. See Kubrin, "Newton and the Cyclical Cosmos," 325–346; and Michael Hoskin, "Newton, Providence and the Universe of Stars," *Journal for the History of Astronomy* 8 (1978), 77–101.

36. On Laplace's theory, and its history and significance, see Stanley J. Jaki, *Planets and Planetarians: A History of Theories of the Origin of Planetary Systems* (New York: John Wiley, 1977), and Ronald L. Numbers, *Creation by Natural Law: Laplace's Nebular Hypothesis in American Thought* (Seattle: University of Washington Press, 1977).

37. Michael A. Hoskin, *William Herschel and the Construction of the Heavens* (London: Oldbourne, 1963).

38. Erich Robert Paul, *The Milky Way Galaxy and Statistical Cosmology, 1890–1924* (forthcoming from Cambridge University Press).

39. E. Robert Paul, "The Death of a Research Programme: Kapteyn and the Dutch Astronomical Community," *Journal for the History of Astronomy* 12, no. 2 (1981), 77–94.

40. After 1930, following the discovery of 'interstellar absorption,' these numbers were scaled down by a factor of three to 100,000 and 30,000 light-years, respectively; see Paul, *The Milky Way Galaxy*, chap. 8.

41. Robert Smith, *The Expanding Universe: Astronomy's 'Great Debate,' 1900–1931* (New York: Cambridge University Press, 1982).

42. See Stephen W. Hawking, *A Brief History of Time: From the Big Bang to Black Holes* (New York: Bantam, 1988), and Eric Chaisson, *Relatively Speaking: Relativity, Black Holes, and the Fate of the Universe* (New York: W. W. Norton, 1987).

43. For their discovery, Penzias and Wilson were awarded the 1978 Nobel Prize for physics.

44. On Einstein and the emergence of relativity theory, see Loyd S. Swenson, Jr., *Genesis of Relativity: Einstein in Context* (New York: Burt Franklin, 1979); Gerald Holton and Yehuda Elkana, eds., *Albert Einstein: Historical and Cultural Perspectives* (Princeton: Princeton University Press, 1982); and A. P. French, ed., *Einstein: A Centenary Volume* (Cambridge, Mass.: Harvard University Press, 1979).

45. J. L. Heilbron, *The Dilemmas of an Upright Man: Max Planck as Spokesman for German Science* (Berkeley: University of California Press, 1986), and Thomas S. Kuhn, *Black-Body Theory and the Quantum Discontinuity, 1894–1912* (Oxford: Oxford University Press, 1978).

46. For a discussion of the development and philosophical meaning of quantum mechanics, see William H. Cropper, *The Quantum Physicists: An Introduction to Their Physics* (New York: Oxford University Press, 1970), and Victor Guillemin, *The Story of Quantum Mechanics* (New York: Scribners, 1968).

47. Niels Bohr, *Atomic Theory and the Description of Nature* (Cambridge: Cambridge University Press, 1934), 52. The sequel to this classic text is idem, *Atomic Physics and Human Knowledge* (New York: Wiley, 1958).

48. Not only is Descartes's metaphysical dualism fundamentally violated, but these developments are reminiscent of Berkeley's doctrine that physical nature is nothing other than ideas (perceptions), apart from which it does not exist at all. Because the Copenhagen interpretation of quantum mechanics has to account for the measuring instruments as physical, material things, it cannot be extended quite so far. For a discussion of this problem, see John D. Barrow and Frank J. Tipler, *The Anthropic Cosmological Principle* (New York: Oxford University Press, 1986), 155.

49. Stephen G. Brush, *The Kind of Motion We Call Heat: A History of the Kinetic Theory of Gases in the Nineteenth Century*, vols. 1 and 2 (Amsterdam: North-Holland, 1976).

50. For a readable account of relativity and quantum theory by one of its founders, see Max Born, *Physics in My Generation* (New York: Springer-Verlag, 1969). For a contemporary account of these developments, see Gary Zukav, *The Dancing Wu Li Masters: An Overview of the New Physics* (New York: William Morrow, 1979).

# 3

## The Nature of Modern Science

While the factual dimension of science is very large, while science can explain a wide range of natural phenomena, while science enjoys enormous status and possesses great authority in the modern world, science is still not a body of answers. Rather, science is a process, a way of asking questions. It is an extraordinarily dynamic process of engaging the world in dialogue with itself and the human mind. If asked the question, Where is science? many would reply that science is in a textbook or in the laboratory. But these perceptions entirely miss the heart of the scientific enterprise: science is a human enterprise, and therefore science resides in the minds of human beings. As a consequence, science is shaped profoundly by human values, by human history, by the philosophical and, at times, even the religious and mystical presuppositions of the milieu in which science is practiced.[1]

### Facts, Theories, and Problem Solving

As an intellectual enterprise, science deals with a range of claims, however tentative initially, from hunches to hypotheses to empirical laws to theories to "laws" of nature. Although this "ordering" might incline one to think of these various levels of claims as ranging along a continuum, this is not the case. For example, empirical laws are regularities that appear in (empirically obtained) data, such as Kepler's three laws of planetary motion or Mendel's laws of inheritance. Laws of nature are idealized abstractions (intellectual constructs) of the way humans understand nature to operate, such as the principle of inertia (Galileo and Descartes) or the first law of thermodynamics (Mayer's and Helmholtz's conservation of energy). Although hypotheses are constantly being proposed as possible explanations, very few hypotheses survive the onslaught of scientific scrutiny to gain full status as theories or general laws of nature. Scientists are constantly inventing new explanations to account for some phenomenon of nature. If the explanation (which begins as a hypothesis) survives severe criticism from the scientific community, then it may achieve status as a theory.

Theories, however, have much higher epistemological status than scientific facts. Except for the most obvious facts (such as "I exist!"), theories as explanations allow us to identify what the facts are. That is, without explanations (theories in a restricted sense) it is usually not possible to isolate the relevant facts from the maze of data in our conceptual environment. The same critique may be applied to all interpretive endeavors, such as anthropology and history as well as natural science and religion.

The term *theory* in the phrase "a scientific theory" means something utterly different than the use of the term *theory* in colloquial conversation, when someone says, for instance, "Well, I have a theory about thus-and-so." In the latter case, the word *theory* refers to a clever idea, a hunch, a rough guess. Frequently people respond, "Oh, that's simply a theory!" or "Oh, that's your theory!" meaning that your hunch is simply your opinion, no more true than someone else's opinion. In contrast, when the scientist says, "The theory of relativity has been strongly confirmed" or "The theory of evolution works," the scientist means that the theory is "true," that it has been strongly confirmed by a considerable amount of evidence.

A scientific theory also conforms to a much larger set of imperatives. Thus, science is not simply composed of, say, physics, biology, and anthropology. Each science is really composed of a group of sometimes overlapping research traditions[2] or paradigms.[3] Thus we can speak of physics as being—at different times—Cartesian, Newtonian, Einsteinian. Each of these research traditions assumes a different ontology of nature and possibly different methodologies. If there is a thread common to all the sciences (or research traditions), it is that they all deal in problem solving. As such, there are a variety of problems—empirical, conceptual, and methodological—that all the sciences engage at various levels of discussion. Within any particular research tradition, the purpose of theories is to explain the factual content of science by reducing all its problems to the ontology of the research tradition.

As with the invention of hypotheses and theories, the solutions to problems are invented by humans, and consequently there is always a context of discovery—a variety of conditions that prevail at the time the discovery is made. For instance, when the plurality-of-worlds idea was first seriously entertained, religious, philosophical, mystical, as well as scientific ideas were all influential in its discovery. The context of discovery is thus very often broad and complex. Once the scientific idea has been discovered, modern scientists nearly always try to justify the idea on the basis of normative scientific principles, which may have nothing to do with its context of discovery.

## Scientific Methods

In the process of justification, science has developed a variety of methods. Frequently, one hears of the "scientific method," as though there is only one method in science. The fact of the matter is that there are a variety of scientific methods, each appropriate to its own discipline(s). The classical scientific method, known as the 'hypothetico-deductive' or H-D method of inference, is used primarily in experimental sciences. The H-D method consists of four steps: (1) the invention of an hypothesis to account for a particular phenomenon (there may be several or many initial possibilities); (2) the deduction of a consequence of the hypothesis; (3) the setting up of some experiment(s) to see if that consequence can be observed; and (4) if the consequence is observed, then, by inference, the confirmation of the hypothesis, and, depending on its empirical and theoretical support, its eventual acceptance as a theory. In this entire process, explanation and prediction are critical components. If the consequence is not experimentally verified, then one of several conclusions results: (1) the hypothesis is simply wrong; (2) the experiment (including the technology needed to test the hypothesis) was ill-conceived or flawed; and/or (3) additional assumptions or restrictions are needed to clarify the hypothesis or experiment. In fact these several possibilities suggest the basic problem with hypothetico-deductive inference—that it always leaves us with an embarrassing superabundance of initial hypotheses.[4]

In addition to the H-D method, there are other scientific methods appropriate to various scientific disciplines. Historically, natural history (the biological and some earth sciences) and some areas of astronomy use descriptive and typological methods. The mathematical sciences (including theoretical physics and astrophysics) use a severe deductive method. More recently, probability and statistical ideas have increasingly become powerful methods within a very large number of the sciences. All scientific fields employ general methods of induction and deduction. General statements or universal regularities are inferred (induced) from less general propositions, while observational claims are deduced from more general views.

In the "experimental" sciences (as opposed to the "mathematical," deductive sciences) the fundamental epistemological problem—and perhaps the most fundamental problem in all the sciences—is that of induction. First rigorously discussed by the eighteenth-century British philosopher David Hume, the problem of induction is this. Suppose we have drawn a number of black-colored balls from an urn. Given that all the balls hitherto drawn were black, does that observed fact constitute sound evidence for the conclusion that all of the remaining, unobserved balls in the urn are black? In other words, would we be justified in accepting that conclusion on the basis of the facts alleged to be evidence for it?[5] The answer to this question has occupied philosophers and scientists for three hundred years. Although many solutions have been offered, including constructivism, logical discovery, falsification and corroboration, and statistical inferential schemes, none have fully satisfied the critics.[6]

Generally, scientists engage their science by attempting to solve problems using methods appropriate to their discipline. In the process of assessing the plausibility of scientific claims, ultimately scientists are involved in falsifying those claims. The degree to which the claims cannot be falsified indicates the degree to which the claims are corroborated and gain scientific authority. Although scientists are not primarily engaged in the work of falsification, for a claim to be strictly scientific, it must in principle be susceptible to experimental scrutiny and therefore falsifiable.

Since many religious, philosophical, and historical claims are not falsifiable, they are by this definition not scientific either. Historically, however, there have been countless scientific claims made that, at the time of their discovery, also did not in fact accord with this methodological imperative. For example, Copernican heliocentrism, Darwinian evolution, mechanistic biology, Einsteinian relativity, Freudian psychology, as well as Newtonian matter theory, Cartesian plenism, Faradayian field theory, and many others that historians could cite made claims or held implied assumptions that at the time were beyond the reach of the science of the day.

## Truth and Scientific Theories

Questions of the truth claims of science are constantly associated with the scientific enterprise. What, in other words, is the status of scientific theories—when are they "true" or, indeed, can they ever be "true"?[7] Since scientists and philosophers have pondered this question, a number of possibilities have been proposed: (1) positivism—the view that theories only summarize facts and that theories per se, while they have no ontological or epistemological status except to organize the factual core of science, can produce "correct" descriptions of reality; (2) instrumentalism—the view that theories are only tools or instruments needed to bring coherence to phenomena; (3) idealism—the view that theories represent only mental constructs and have no ontological value; and (4) realism—the view that theories have a direct correspondence to the world and that science, therefore, is a correct description of reality.

Questions of truth in science all presuppose a naive objectivism that ignores the inevitable role of our own concepts in our accounts of reality.[8] For this reason, to a degree science is bound to its own historical space and time, reflecting cultural biases in its choice of metaphysical assumptions. In order to be able to distinguish science from other human endeavors, such as art, music, and religion, however, we must be willing to acknowledge the orientation of science toward truth.

Still, one must look elsewhere to the idea of a "theory" for the truth adequacy of scientific claims. Let us suppose that a particular scientific claim has achieved the status of theory. This means (1) that there is substantial empirical, experimental, and/or mathematical support (evidence) for the theory; (2) that there are no serious anomalies remaining that the theory cannot explain; (3) that, except for a very few, the scientific community at large has achieved consensus on this theory; and (4) that the theory is part of a much larger conceptual structure and fits coherently into that larger frame. Thus, for example, the theory for hominid evolution is part of the much larger theory of evolution, which in turn is subsumed in the larger superstructure of research tradition or paradigm. Whether research tradition or paradigm, however, all superstructures make fundamental methodological and ontological assumptions. Thus, for example, in the process of totally rejecting Creationism (note the capital *C*), modern evolutionary theory assumes an ontological world that rejects the following: essentialism (the idea that a species has preexistent status), nominalism (that the species idea has no special status), anthropocentrism (that the world, at the biological level, is human-oriented), and creationism (that God created all species at the beginning without any possibility of phylogenetic change).[9]

The point is that, by definition, for a theory to have any scientific status it must be subsumed in a research tradition, which in turn is a "set of ontological and methodological do's and don'ts."[10] The purpose of theories in research traditions then becomes to reduce the empirical problems to the ontological and methodological requirements of the research tradition. Consequently, science never, in some ultimate sense, makes truth claims with a capital *T*—rather, science is a human process that allows humans to build conceptual structures.

This contrasts with the rhetoric of many scientists, who, both historically and recently, have assumed that science fundamentally is not concerned whether nature should be one way or another. Nature, their argument goes, is simply value-free and completely neutral; therefore science, as the reflection of nature, will only be descriptive and true—never prescriptive and tentative. The argument that science is utterly neutral and value-free is grounded in the assumption that there is a one-to-one correspondence between science and an "objective" reality "out there" upon which science operates. The history of science, particularly in recent decades, has shown conclusively that this assumption is false. The truth of the matter, as we have seen, is that science does reflect its cultural matrix.[11] If there were some deep (and provable) synonymous relationship between science, on the one hand, and reality, on the other, then, for example, Aristotelianism, Newtonianism, and Cartesianism should all be true. But then so should all the other countless scientific theories and research traditions that have been shown to be inadequate by the standards of the day. The issue is not whether the empirical facts one observes are true; to the degree that those facts are observed, one might correctly contend that they are true. Rather, the issue deals with the epistemological status of the explanation of those facts—namely the theories. To this end, it is correct to say that science seeks to construct conceptual models of perceived realities. These models, however, reside in the minds of human beings; there is no persuasive evidence that they do not have an existence independent from their creators.

In the final analysis, the debate whether scientific claims ultimately reflect Platonic pure ideas (mathematical essentialism) or whether they are Kantian human constructs (mathematical modelism) is moot at best. We do know that science resides in the minds of human beings. And to that degree science is a construct, a model, an abstract conceptual structure of—or about—reality. In short, we should never make the mistake of confusing science with reality. Science as science is NOT reality! Nor, for that matter, is theology or religion reality. All of these human enterprises are just that: human! They are ways by which humans organize their understanding of reality; by themselves they are not reality itself. They are what we might call 'meta' structures. So the question becomes, How closely do these 'meta' structures approximate reality? We can only ascertain the ontological status of some scientific or religious idea if that idea is given from God—directly by revelation. Unfortunately, if some human interprets the revelation, it must be done in human terms (language), and so doing removes the original revelation one more step from those not privy to that original theophany.

Thus some philosophers now talk not of science as truth but of science converging toward some objective reality.[12] Known as 'convergent realism', this view is still philosophically insecure. Even though it argues for a weak form of truth—namely "convergence"—by definition it can never provide criteria that will allow us to assess if and when the truth is obtained. In short, it really begs the issue; by definition the alleged convergence must always remain beyond our grasp. Therefore any talk of convergence simply avoids the inevitable—what was known when we started.

Occasionally it is suggested that science is science only when it measures and quantifies. Anything else is speculative and, at worst, metaphysical. This claim is only partly true, however. Not only are there sciences that are by their very nature purely conceptual, such as theoretical physics and certainly most of modern mathematics, but, even more vital, the quantification of nature by itself tells us nothing without the larger, theoretical framework needed to give shape to the measurable and quantifiable. For example, the essential difference between the geocentric and the heliocentric systems is perspective. Copernicus, Kepler, Galileo, and a host of others provided no measurable data that compelled people to think heliocentrically. To be sure, their further investigations eventually helped to confirm the heliocentric system. But even without the data one cannot doubt that Copernicus's heliocentric hypothesis was science. In science the conceptual framework is the sine qua non. No respectable scientist, philosopher, or historian of science would contend, for example, that Einstein's theories of special and general relativity were not science prior to their empirical verification. Whether their work is primarily conceptual or experimental, when scientists attempt to understand the workings of nature, they are engaged in science as the pursuit of truth, however illusive.

Fundamentally, the problem in objectifying the truth claims of science requires some sort of neutral criteria—a language or metascience. But historically all such positivistic approaches have failed. No other human endeavor has this unique characteristic, yet the scientific enterprise is singular in its ability to achieve disciplinary consensus by applying a variety of criteria to the assessment of its tentative claims. Consequently, although "truth" is forever beyond our grasp, theories and supertheories (research traditions or paradigms) can achieve the appearance of truth, albeit tentative truth.

As an intellectual enterprise, science is basically an endeavor that attempts to build a mental construct of human perceptions. Science, therefore, should never be confused with reality itself. The immediate consequence of this is that science, by its very nature, must change. Therefore, constructing scientific theories is not dissimilar to nailing jello to a wall: the moment the theory has been fashioned, it begins to lose its conceptual grip as scientists undertake the process of careful and detailed scrutiny. The truth claims of a scientific theory are measured in terms of the theory's capacity to explain a phenomenon and predict its consequences. Therefore, those theories that do survive do not have superior epistemological status; rather, more properly, they have superior explanatory and predictive capability. In the final analysis, a theory's survivability is measured in terms of its capacity to reduce problems to the ontological and methodological norms of its research tradition.

## Science, Religion, and Foundational Assumptions

The ultimate accommodation of science with religion must recognize that both science and religion rest on crucial methodogical and metaphysical assumptions. In science, these are: (1) universality, known as uniformitarianism and actualism—the view that the universe operates temporally in the same fashion and the same degree, respectively, yesterday, today, and tomorrow, as well as spatially and isotropically in all directions and places; (2) simplicity (sometimes called parsimony or 'Ockham's Razor')—the assumption that, given alternative ways of explaining natural phenomena, nature operates in the simplest of all possible ways; (3) naturalism—the view that nature operates without extraphysical influences; (4) rationalism—the view that the universe is rationally ordered and understandable by rational means (this has remained a sine qua non of modern science since its founding in the seventeenth century); and (5) coherence—the methodological imperative that, across the entire scientific matrix, science must strive for internal compatibility and consistency; in other words, the "truth" in one scientific discipline must not be contradicted by results in another scientific field.

One must not make the mistake, however, of assuming that these scientific foundational assumptions are merely assumptions. The extraordinary coherence of the scientific enterprise is astounding. If, for example, universality were not a genuine characteristic of the universe, science would be reduced to a collection of disconnected claims and conclusions, no one of which would possess any more epistemological status than any other mere guess. The historical and contemporary evidence for this view (really no longer an assumption) is overwhelming.[13] On the other hand, choosing the simplist of possible explanations does not in itself guarantee a correct, valid, or true explanation.

Religion also makes crucial methodological and metaphysical assumptions about the nature of reality and its knowability. In Mormonism, for our purposes these are: (1) there exists a God who is conscious and actively involved in His creation; (2) humans are the direct and literal offspring of God; (3) law, both natural and moral, is a creation of, with, or before God; (4) but God can and occasionally does contravene natural and moral law (miracles); and (5) humans can know God, and of Him, through extraphysical, transcendent means, such as revelation, inspiration, and intuition.

The degree to which science and religion are perceived as being ontologically compatible hinges crucially on the status of these foundational assumptions. If these sets of assumptions are perfectly compatible, then religion and science would stand on the same ground and therefore should also be fully commensurate. They are not, however. For example, in ultimate terms, Mormonism and science make radically different methodological presuppositions. Moreover, in science there can be no violation of law, as allowed by religion.

Still, there is a middle ground that allows for a compatibility of science, religion, and truth. That ground, however, rests crucially on the very tentative nature of science vis-à-vis its relationship to ultimate reality. Humans exist in the 'phenomenal' world. God and transcendent reality, including law (assuming the latter exists apart from the physical conditions of the universe), exist in an ultimate or 'noumenal' world. Though a human activity, science attempts to determine the operation of the phenomenal world by inferring the nature of the noumenal world. That inference is a conceptual world, a human construct that reflects the nature of the noumenal but itself is never a perfect reflection of the noumenal. It may not even be a slightly accurate correspondence, because there is no way of validating it with certainty. Consequently, this human, conceptual world must remain slippery and forever changing.

The theistic scientific view—that true science and true religion will never conflict—is based on the assumption that humans may have developed a conceptual world that is identical to the noumenal world of ultimate reality. Whether that correspondence is one-to-one or not is the fundamental issue. As I have argued, although there are deep connections, at best one can only make the assumption that there is a correspondence between the two—and then one needs to explore the degree of that correspondence. As we have seen, to consider the degree of correspondence is almost a meaningless question, because the history of science has provided incontrovertible evidence that science as a conceptual, human enterprise is forever changing. To answer the question of the degree of correspondence philosophically would require either (1) transcendent knowledge, hardly forthcoming except in the most general terms, (2) an argument based on science itself, which therefore becomes tautological and hence logically unacceptable, or (3) the invention of a metalanguage, the construction of which, given the history of such philosopher-scientists as Ludwig Wittgenstein, Bertrand Russell and Alfred North Whitehead, and David Hilbert, has been shown to be impossible, except with the most elementary propositions.[14]

## NOTES

1. See I. Bernard Cohen, *Revolutions in Science* (Cambridge, Mass.: Harvard University Press, 1985), and Gerald Holton and R. S. Morison, eds., *Limits of Scientific Inquiry* (New York: W. W. Norton, 1979).

2. See Larry Laudan, *Progress and Its Problems: Towards a Theory of Scientific Growth* (Berkeley: University of California Press, 1978); idem, *Science and Values* (Berkeley: University of California Press, 1984); and Arthur Donovan, Larry Laudan, and Rachael Laudan, eds., *Scrutinizing Science: Empirical Studies of Scientific Change* (Boston: Kluwer Academic Publishers, 1988).

3. Thomas S. Kuhn, *The Structure of Scientific Revolutions*, 2d ed. (Chicago: University of Chicago Press, 1970); idem, *The Essential Tension: Selected Studies in Scientific Tradition and Change* (Chicago: University of Chicago Press, 1977); and Gary Gutting, ed., *Paradigms and Revolutions: Applications and Appraisals of Thomas Kuhn's Philosophy of Science* (Notre Dame: University of Notre Dame Press, 1980).

4. Wesley C. Salmon, *The Foundations of Scientific Inference* (Pittsburgh: University of Pittsburgh Press, 1967), 108–32.

5. Ibid., 7.

6. See Carl Hempel, *Aspects of Scientific Explanation and Other Essays in the Philosophy of Science* (New York: Free Press, 1965); Norwood R. Hanson, *Patterns of Discovery: An Inquiry into the Conceptual Foundations of Science* (New York: Cambridge University Press, 1969); Karl Popper, *The Logic of Scientific Discovery* (New York: Harper and Row, 1959); idem, *Conjectures and Refutations: The Growth of Scientific Knowledge* (New York: Basic Books, 1962); and Salmon, *The Foundations of Scientific Inference*.

7. On the nature of theories, see Frederick Suppe, ed., *The Structure of Scientific Theories*, 2d ed. (Urbana: University of Illinois Press, 1977).

8. See, for example, Arthur I. Miller, *Imagery in Scientific Thought: Creating Twentieth-Century Physics* (Cambridge, Mass.: MIT Press, 1986); Gerald Holton, *The Advancement of Science, and its Burdens* (New York: Cambridge University Press, 1986); idem, *The Scientific Imagination: Case Studies* (New York: Cambridge University Press, 1978); and idem, *Thematic Origins of Scientific Thought: Kepler to Einstein* (Cambridge, Mass.: Harvard University Press, 1973).

9. Some contemporary creationists accept the possibility of phylogenetic change at the level of species, but they deny that this has any implications for differences at higher levels of organization.

10. Laudan, *Progress and Its Problems*, 80.

11. Not only has the history of science in recent decades overwhelmingly demonstrated the validity of this proposition, but many scientists and philosophers of science in recent years have begun to recognize its legitimacy. Among numerous scientists we could note, see the works cited above by Gerald Holton, R. S. Morison, and Arthur Miller, and Ernst Mayr, *The Growth of Biological Thought: Diversity, Evolution, and Inheritance* (Cambridge, Mass.: Belknap Press, 1982). Among philosophers of science, also see the works cited above by Thomas S. Kuhn, Larry Laudan, I. Bernard Cohen, as well as Edwin Arthur Burtt, *The Metaphysical Foundations of Modern Physical Science*, rev. ed. (New York: Humanities Press, 1951); Norwood R. Hanson, *Patterns of Discovery: An Inquiry into the Conceptual Foundations of Science* (New York: Cambridge University Press, 1969); Stephen Toulmin, *Human Understanding* (Oxford: Clarendon Press, 1972); Wolfgang Stegmueller, *The Structure and Dynamics of Theories* (Berlin: Springer-Verlag, 1976); Imre Lakatos, *Philosophical Papers*, Vol. 1. *The Methodology of Scientific Research Programmes* (New York: Cambridge University Press, 1978); and Paul Feyerabend, *Rationalism, Realism and Scientific Method: Philosophical Papers*, Vol. I (New York: Cambridge University Press, 1981).

12. See Jarrett Leplin, ed., *Scientific Realism* (Berkeley: University of California Press, 1984), and Colin Howson, ed., *Method and Appraisal in the Physical Sciences: The Critical Background to Modern Science, 1800–1905* (New York: Cambridge University Press, 1976).

13. For an assessable discussion of this view, see James Trefil, *Reading the Mind of God: In Search of the Principle of Universality* (New York: Anchor Doubleday, 1989). For a contemporary perspective on this issue by a Mormon geologist, see Jess R. Bushman, "Hutton's Uniformitarianism," *Brigham Young University Studies* 23, no. 1 (1983), 41–48.

14. The question of the epistemological status of science and mathematics has been one of the central issues in philosophy, particularly during the first several decades of the twentieth century. See, for example, Wittgenstein's linguistic analysis in his *Tractatus Logico-Philosophicus*, trans. D. F. Pears and B. F. McGuiness (London: Routledge & Kegan Paul, 1961); Whitehead and Russell's logicist efforts in their *Principia Mathematica* (Cambridge, England:

University Press, 1910), and Hilbert's mathematical formalism program, as in his *Grundlagen der Geometrie* (1898) and his *Grundlagen der Mathematik* (Berlin: Springer, 1968–70). For an extremely insightful discussion of these issues in their cultural context, see Allan Janik and Stephen Toulmin, *Wittgenstein's Vienna* (New York: Simon and Schuster, 1973).

# PART TWO

# Mormonism and Cosmology

# 4

## Joseph Smith and Cosmology

Theologically, Joseph Smith presented a cosmology relating man, God, and the universe; religiously, his message was millenarian and directed toward eschatological issues; and, philosophically, his cosmic vision encompassed an astronomical pluralism that provided for the quintessential understanding of postmortal existence. Joseph Smith did not, however, consciously set out to develop any of these themes to their fullest. To flush out relevant origins of Joseph Smith's private cosmic vision, one must explore the interplay of his vision with its most concrete manifestation—namely, the doctrine of 'worlds without number.' Through the pluralist dogma it is possible to grasp the full scope of both his theological and religious ideas, because for Joseph Smith these ideas are ultimately justified in his vision of postmortal humankind. And that development is most clearly presented in his universal vision, which encompasses all of creation.

Joseph Smith's almost unwitting excursion into theological and metaphysical issues thrust him directly into cosmology—his private vision of the cosmos that ultimately dictated his understanding of the origin and structure of the universe. The cosmic vision Joseph Smith adopted thus moved him into the 'plurality of worlds' tradition developed, as we have seen, in the seventeenth century. In the eighteenth century Protestant evangelicals began using the plurality-of-worlds idea in natural theology—the view that nature contains clear and compelling evidence of God's existence—to substantiate their own beliefs in Christianity and to counter threats of religious skepticism. Joseph Smith's use of the concept was radically different, though never defensive, because ultimately it was integrated into and defined the boundaries of his unique cosmic vision.[1] Even though Joseph Smith used the plurality-of-worlds idea extensively, he said almost nothing about science, scientific cosmology, or the relationship of science to religion. Thus what follows focuses primarily on his plurality doctrine as a vehicle for understanding an innovative and complex cosmology.

### Deism and the Plurality of Worlds

The plurality-of-worlds doctrine accompanied the birth of modern science in the seventeenth century—a time that fostered the growth of natural theology because scientific and religious views complemented mutual intellectual concerns. By the beginning years of the nineteenth century, the concept of multiple inhabited worlds found wide acceptance among secularists, deists, and theologians. The early nineteenth-century Scottish evangelicals Thomas Chalmers, Thomas Dick, and later David Brewster and Hugh Miller all wrote on the plurality of worlds and stressed the compatibility of science and religious beliefs. Particularly in the context of Anglo-American developments, science increasingly supported the structure of biblical understanding. Not only was God's word a testament of His continuing interests in human affairs, but His *works* gave abundant evidence of the nature, power, and majesty of the divine presence. The secular tradition of plurality of worlds was transformed into a religious idea and frequently seen as an endeavor to support religious views. Ironically, however, the plurality-of-worlds concept was used not only by sectarians to substantiate their faith but also by deists and others to debunk the claims of Christianity. Thus both Christians and secularists, believers and deists found in the pluralist doctrine evidence to support their views. As a result, pluralism filtered throughout American frontier society, not only in the writings of such popular figures as the evangelicals Chalmers and Dick but also with deists such as Thomas Paine. Pluralism was disseminated by books, newspapers, and almanacs, as well as orally, at religious gatherings and casual meetings.

The most widely known source of deism in early nineteenth-century America was Tom Paine's *Age of Reason* (1794). Fanned by the fires of the French Revolution and the Enlightenment, Paine popularized along the lines of English anticlerical and rationalistic thinking. Assuming the apotheosis of humankind's divine gift of reason, Paine's thesis was that Christian theology is fundamentally incompatible with both human reason and humankind's increasingly scientific understanding of the universe. Reason alone, and not biblical myth, is capable of informing humans of the universe and its laws of operation.

Paine, who had gradually immersed himself in the new astronomy of the seventeenth and eighteenth centuries, made his understanding of science—and astronomy in particular—the basis of his deism. Part 1 of *The Age of Reason* manifests the power of astronomy over Paine and the central position the plurality of worlds came to occupy in his theological and scientific thinking. Thus, after describing in detail the immensity of the solar system, Paine extended his views to the vastness of the cosmos:

> Beyond this [the solar system], at a vast distance into space, far beyond all power of calculation, are the [fixed] stars. . . . Those fixed stars continued always at the same distance from each other, and always in the same place, as the sun does in the center of our system. The probability, therefore, is that each of those fixed stars is also a sun, round which another system of worlds or planets . . . performs its revolutions, as our system of worlds does round our central sun. . . . [T]he immensity of space will appear to us to be filled with systems of worlds, and that no part of space lies at waste.[2]

Intellectual historian Marjorie Hope Nicolson has argued that "the real basis of Paine's 'deism', . . . the chief source of his theological beliefs . . . is the climatic and inevitable popularizing of . . . the controversy whether ours is not merely one of a plurality—even, some dared to think, of an infinity—of worlds, and whether such of these universes may not possess rational inhabitants."[3] On the basis of his understanding of astronomy, Paine believed that every evidence of science either "directly contradicts the Christian system of faith or renders it absurd."[4] Thus, he wrote:

> From whence, then, could arise the solitary and strange conceit that the Almighty, who had millions of worlds equally dependent on his protection, should quit the care of all the rest and come to die in our world, because, they say, one man and one woman had eaten an apple? And, on the other hand, are we to suppose that every world in the boundless creation had an Eve, an apple, a serpent, and a redeemer? In this case, the person who is irreverently called the Son of God, and sometimes God himself, would have nothing else to do than to travel from world to world, in an endless succession of deaths, with scarcely a momentary interval of life.[5]

Paine's unrelenting attack on Christianity, the support he marshalled for his views in terms of the plurality of worlds and astronomy, and his claims of deism generally all entered the popular culture of early nineteenth-century American thought. As the American historian Merle Curti has pointed out, "Humble men in villages from New Hampshire to Georgia and beyond the Alleghenies discussed it by tavern candlelight."[6] In the six years between its publication and 1800, at least sixteen published criticisms of *The Age of Reason* appeared. The British Museum Catalogue lists more than fifty published responses.[7]

Although many wrote responses to Paine's *Age of Reason*, others used it with missionary zeal to combat their perceptions of religious tyranny. Evidence indicates that even Joseph Smith's father possessed a copy. Joseph Senior's father, Asael, disapproved of Methodism, perhaps "because of its vigorous preaching of the eternal condemnation of the unregenerate," a view in contrast to Asael's own universalism.[8] Consequently, according to Lucy Mack Smith's unpublished manuscript, when his son was considering joining the Methodists, Asael "came to the door one day and threw Tom Paine's *Age of Reason* into the house and angrily bade him read it until he believed it."[9]

Although it has been argued that Joseph Smith himself read *The Age of Reason* and wrote the Book of Mormon in defense of Christianity, there is no evidence that he either possessed or read a copy of Paine's work. It seems almost certain that he was acquainted with the ideas of 'natural religion' but shared New England attitudes that were, by the 1820s, already strongly and consciously opposed to infidelity.[10] In his insightful study of Joseph Smith, Richard Bushman has argued that the Smiths had been "more directly affected by Enlightenment skepticism than by Calvinist evangelism" and thus "were destined to live along the margins of evangelical religion."[11] Though skepticism lost ground for a time early in the nineteenth century, it revived beginning in the 1820s with the founding of various periodicals and remained an influence in frontier villages through newspaper editors and other homespun intellectuals.[12] Thus, as far as deism is concerned, it is irrelevant whether Joseph Smith was acquainted personally with *The Age of Reason*. Also, Paine's plurality-of-worlds concept was *not* presented in a manner that would have aided Joseph Smith in developing his own complex system of pluralist concepts. In fact, Joseph Smith's theology of astronomical pluralism would more easily have been developed from sources other than Paine's *Age of Reason*. We can be certain, though, that the plurality-of-worlds concept was widely understood and diffused during the early years of the century, partially as a direct consequence of Tom Paine.

## Religious Orthodoxy and Astronomical Pluralism

By far the most influential source of astronomical pluralism in the 1820s, however, was no longer *The Age of Reason* but *A Series of Discourses on the Christian Revelation Viewed in Connection with the Modern Astronomy* (1817) by the Reverend Thomas Chalmers, a young, gifted Scottish minister. Though other writers, such as the English poets Shelley and Byron, espoused pluralism in the first decades after Paine's work, it was the American edition of Chalmers's *Astronomical Discourses* that introduced the American reading public to the unparalleled appeal of the magnificence of God's creation. Originally delivered as a series of seven complementary religious sermons, *Astronomical Discourses* met with instant and wide success. Its influence, like that of *The Age of Reason,* was truly astounding and prompted the Reverend Edward Hitchcock four decades later to express the view that "all the world is acquainted with Dr. Chalmers' splendid *Astronomical Discourses.*"[13]

Besides displaying Chalmers's brilliant literary talents, the work appeared in America at a time characterized by religious revivals and evangelical fervor, when deism and rationalism were increasingly associated with infidelity and the excesses of French revolutionary tyranny and Jacobin extremism.[14] Furthermore, it was in *Astronomical Discourses* that *The Age of Reason* finally found a worthy opponent. As Paine had earlier used the plurality of worlds to argue against Christianity, Chalmers now used pluralism to support and defend the revival of religious neo-orthodoxy. Chalmers's intention thus became twofold: (1) to counter the skeptics' arguments and to remove difficulties in the way of belief, and (2) to examine the implications for Christian belief entailed in science and astronomy, particularly as suggested in the plurality doctrine.[15] Thus Chalmers's *Astronomical Discourses* became a significant example of natural theology—science in the service of one's religious convictions.

Paine's most serious criticism of Christianity dealt with the presumed absurdity that Jesus Christ, the incarnation of God Almighty, should either come to this Earth to extend the atonement and redemption to *all* his creations or travel from world to world in an endless succession of deaths. Paraphrasing the infidel, Chalmers introduced his first sermon by citing from the Psalms that "he is mindful of us" and then stating the main theme of the discourses:

> This very reflection of the Psalmist has been appropriated to the use of infidelity, and the very language of the text has been made to bear an application of hostility to the faith. "What is man that God should be mindful of him or the son of man, that he should deign to visit him?" Is it likely, says the Infidel, that God would send His eternal Son, to die for the puny occupiers of so insignificant a province in the mighty field of His creation?[16]

In one form or another, this theme thoroughly dominates the *Astronomical Discourse.* Because of the relevance of Chalmers's treatment of deism, as well as skepticism and the pluralist doctrine and Joseph Smith's treatment of these issues, it may be useful to summarize Chalmers.

After introducing his main theme, Chalmers first sketches the dimensions of our solar system and the extent of the stellar universe, including the idea of multiple inhabited world systems. Here, Chalmers argues, because God's benevolence extends to all his creations, including the most insignificant of creatures, Christians should not be disturbed by God having sent "His eternal Son, to die for the puny occupiers of so insignificant a province in the mighty field of His creation."[17] In the second discourse, entitled "The Modesty of True Science," Chalmers praises the empiricism of Sir Isaac Newton, vis-à-vis the rationalism of Voltaire, and argues that the skeptic uses selective evidence and criteria based upon unverifiable assertions to speculate on other worlds, while denying the universal applicability of Christianity. "The Extent of the Divine Condescension," the third discourse, challenges the assumptions imposed upon Christianity by the skeptic. Chalmers defies his opponents to indicate a single instance of God's inability to deal with the details of his universe. Moreover, to assume that God lacks commitment to his creations misrepresents the divine presence. In answer to the question of Christ's atonement, "The plan of redemption may have its influences and its bearings on those creatures of God who people other regions, and occupy other fields in the immensity of his dominions; that to argue, therefore, on this plan being instituted for the single benefit of the world we live in, and of the species to which we belong, is a mere presumption of the Infidel himself."[18] Although the scriptures are not intended to give us a knowledge about worlds other than ours, Chalmers suggests in his fourth discourse that, just as Christ's redemption is efficacious through the millennia of human history, so the atonement reaches throughout the universe. Not only is the human drama pursued with intense interest by the angels (the fifth discourse), but, just as a small and perhaps insignificant piece of land may decide the results of larger interests (the sixth discourse), so the Earth, tiny and insignificant as it is, may decide the outcome of struggles between light and darkness universally.

The primary purpose of Chalmers's *Astronomical Discourses* was not to lecture on the plurality of worlds or even to counter Paine's *The Age of Reason* but to awaken its readers to the power of God's saving word. Thus, ending with the seventh discourse, Chalmers reminds his readers that they are agents in a cosmic battle and that Christianity provides not only the knowledge but the power needed for universal and personal salvation. Whether Chalmers succeeded as an evangelist is not entirely certain. What is clear, however, "many [of his readers] left it convinced pluralists, certain that Christianity not only could be reconciled with that doctrine but also could thereby attain new grandeur."[19]

While Chalmers's *Astronomical Discourses* was known widely in America, other sources of pluralism and Christian doctrine were also influential. Almost without exception, these sources are all examples of natural theology, any one of which could have been used equally to support the Christian message. Less accessible but perhaps more important as a source of ideas on astronomical pluralism among ministers was the work of the noted Calvinist Timothy Dwight. As president of Yale University from 1795 until his death in 1817, Dwight delivered 173 sermons, published in 1818 as *Theology Explained*, to Yale undergraduates "to save them from infidelity, to improve their morality, and to instruct them in Christianity."[20] During Dwight's tenure, as many as one-third of Yale undergraduates went on to study for the ministry. As these men fanned throughout New England and the western territories, no doubt many of their sermons asserted, implicitly or otherwise, the pluralist doctrine.[21]

In many of his sermons Dwight drew heavily upon astronomical pluralism and natural theology generally.[22] Like Chalmers, Dwight felt compelled to answer Paine's central criticism of Christianity—that Christ would be forced either to travel from world to world in an endless succession of deaths or to atone on this Earth for all of God's creations:

> This world was created, to become the scene of one great system of Dispensations toward the race of Adam; the scene of their existence, and their trial, of their holiness, or their sin, and their penitence and reformation, or their impenitence and obduracy. It was intended, also, to be a theatre of a mysterious and wonderful scheme of providence. The first rebellion in the Divine Kingdom commenced in Heaven: the second existed here. The first was perpetrated by the highest, the second by the lowest, order of Intelligent creatures. These two are with high probability the only instances, in which the Ruler of all things has been disobeyed by his rational subjects. The Scriptures give us no hint of any other conduct of the same nature: and no beings are exhibited in them as condemned at the final day, or sent down to the world of perdition, beside fallen angels, and fallen men. As, therefore, these are often mentioned as fallen creatures, and these only; it is rationally argued, that no other beings of this character have existed.[23]

As Dwight asserted, Christ's atonement was needed only by the inhabitants on this earth and is therefore unique among God's creations as a limited redemption.

In addition to the pluralism presented in Paine, Chalmers, and Dwight, a fourth widely read source was available in rural America. Fawn Brodie has claimed that by 1835 Joseph Smith had recognized the importance of formulating a metaphysics that would rationalize science with his own special brand of "Jewish and Christian mysticism." That synthesis, she argued, was the book of Abraham, and a major source of ideas was Thomas Dick's *Philosophy of a Future State*, a work first published in 1828 that, Brodie claims, Joseph Smith "had recently been reading" and that "made a lasting impression" on him.[24]

Not only Dick's *Philosophy* but nearly all of his ten books were laced with astronomical pluralism. Though Dick's first work, *The Christian Philosopher; or, The Connection of Science and Philosophy with Religion* (1823), did not specifically deal with the plurality of worlds, it launched him on a successful career as a writer of science, religion, and natural theology. In his *Philosophy of a Future State* (dedicated to Thomas Chalmers), Dick speculates on the plurality of worlds in increasing detail, even calculating the numbers of inhabited worlds within the universe. His approach to the plurality doctrine and science generally assumed a cosmos characterized by purpose, order, and direction. This sort of teleological approach was often developed within natural theology, yet nowhere in his extensive writings does Dick feel compelled, as Chalmers and Dwight did earlier, to answer Paine's objections to Christianity. Even without this defense of the faith, Dick's writings became extremely popular in Britain and America and served to sustain much interest in the pluralist view.

Although Chalmers, Dwight, and Dick all speculated widely on the topic of the plurality of worlds, Joseph Smith's system of cosmology was developed with a metaphysical and theological daring not apparent in their much longer and more developed studies. For instance, Joseph Smith's references to pluralism in time, to hierarchically ordered worlds with their own laws and bounds, and to the specific kinds of inhabitants on these worlds are not developed in these other works. Whereas Chalmers's intention was always directed to evangelical conversion, Dwight's purpose was to enhance ministerial indoctrination, and Dick's hope was to expose the public to a natural theology of Christianity, Joseph Smith developed pluralism into an innovative metaphysical cosmology of postmillennial significance. Though it may be doubtful that Joseph Smith consulted any of these works, it is probable that he heard them discussed in formal or casual conversation. Indeed, we can posit with reasonable confidence that Joseph Smith first heard of the plurality idea during the revivalistic meetings of his youth. Chalmers, Dwight, Dick, and nearly all other evangelicals wrote on the plurality of worlds, and science in general, as an example of natural theology to support Christian evangelicalism.

## Pluralist Thought on the American Frontier

Over the last fifty years, it has been routinely suggested that during the 1820s Joseph Smith made use of the area's most important library.[25] Sometime around 1815, in the township of Farmington just five miles south of the Smith farm, the Manchester Rental Library Society was organized. As one of the region's first libraries open to all patrons who paid for initial membership and continued with annual dues, the Manchester Library included a wide selection of books; the collection eventually grew to at least 421, of which 275 were actually purchased in or before 1830. Included here were copies of Thomas Dick's *Christian Philosopher* (1826 ed.) and his *Philosophy of the Future State* (1829 ed.), and a copy of Andrew Fuller's *Gospel Its Own Witness; or, The Holy Nature, and Divine Harmony of the Christian Religion, Contrasted with the Immorality and Absurdity of Deism* (1803 ed.). Dick's *Philosophy of the Future State* is saturated with pluralism, whereas Fuller, though not nearly as influential as others we have considered, joins with Chalmers and Dwight in refuting Paine's *Age of Reason*. As a critique of deism, Fuller's book was intended more as a religious work than as a defense of pluralism.

Despite the claims of some writers, none of the principals involved in the early years of the Restoration—including Joseph Smith—were members of the Manchester Rental Library Society nor did any of them make use of its rich, though relatively sparse, resources. Moreover, if Joseph Smith had wished to explore the literary materials of the day, it would have been unnecessary to travel the five miles to Manchester when in Palmyra, only two miles distant, there were several bookstores and at least one "library," the contents of which he would presumably have been free to peruse.[26] The contents of these various "libraries," with the exception of the Manchester Rental Library, are no longer preserved.

It is possible to surmise the holdings of the various libraries by examining the lists of books available for purchase in the Palmyra–Manchester–Canandaigua area as advertised in local newspapers.[27] Between 1820 and the early 1830s, the period during which Joseph Smith was active in bringing forth his new scriptural writings, presumably he would have had access to the local newspapers.[28] During this period, these newspapers occasionally advertised for sale "a general and well selected assortment of books," including *The Works of Thomas Chalmers* and Dwight's *Theology Explained*.[29] While they also occasionally carried articles on science, particularly on such astronomical topics as "solar spots," comets, and meteors, they rarely, if ever, discussed the pluralist doctrine explicitly.[30]

Besides books and newspapers, perhaps the most widely read literature was almanacs.[31] Literature, art, historical and current events, manners, morals, and entertainment were often presented in almanacs. In addition to calendric information, eighteenth-century almanacs included astronomy, mathematics, Copernican theory, Newtonian mechanics, natural history, geology, and medicine. Typical astronomical data that might affect the weather—and according to some, humanity itself—included the positions of the sun and moon, the moon's phases, the position of the planets, and dates of eclipses.[32] For many years the philomath almanacs of colonial times emphasized natural philosophy and particularly astronomy, including the plurality of worlds.[33] As farmer's almanacs were developed toward the end of the seventeenth century, they emphasized more utilitarian concerns, and fewer discussions of pluralism emerged in late eighteenth- and nineteenth-century editions.

In Palmyra, as throughout America, in every year from 1818 through 1830 notices advertising almanacs appeared in newspapers. Almanacs were also on sale in bookshops in both Palmyra and Canandaigua, but they rarely discussed science and natural philosophy and, with only a few exceptions, never dealt with astronomy, let alone astronomical pluralism. For example, in Andrew Beers's *Farmer's Diary* (1824), published by James D. Bemis, editor of the *Ontario Repository*, is printed one of the few essays on astronomy found in this period. Entitled "Formation of the Universe," this essay, however, only obliquely assumed the notion of the plurality of worlds. Generally speaking, even though almanacs were widely available, they represented a poor source of ideas dealing with the pluralist doctrine.

Other sources of a *formal* discussion of pluralism are possible, but not likely. Masonic thinking does not consider the plurality of worlds. The most relevant parallel between Mormonism and Freemasonry is the common use of certain astronomical symbols. In the construction of the Nauvoo Temple, for instance, sun, moon, and star stones adorned its exterior. In Freemasonry, these images symbolize degrees of understanding, whereas in Mormon temple cosmology they represent the several heavens of Mormon afterlife with all their pluralistic implications of multiple world systems. Even if Joseph Smith was influenced by Freemasonry in his temple theology (an enduring but problematic thesis), such influence did not extend to his ideas on astronomy and its implied pluralism.[34]

Were other—more ancient—sources dealing with the astronomical cosmology of either Abraham or Enoch available to Joseph Smith? Excluding the canonized scripture of orthodox Christianity, the only Greek writings (in English translation) available in area bookshops and libraries in Palmyra and Canandaigua were editions of Josephus's *Works*. In his discussion of Jewish antiquities, however, Josephus barely touches on Abrahamic astronomy and nowhere discusses astronomy in any significant detail. Elsewhere, however, derivatives of Abrahamic astronomy were considered in the writings of some classical Greek authors. Unfortunately, the only writings of Greek origin advertised in local newspapers and available in bookstores and libraries included an occasional grammar, reader, or New Testament. Even the works of Aristotle, Plato, and the neo-Platonists such as Proclus were rarely found in the area. In fact, the only serious classical source discussing Abrahamic astronomy was Thomas Taylor's 1816 English translation of Proclus's *Theology of Plato*, a work virtually unknown in America at the time.[35] Not until 1840 did a few apocryphal sources, such as the works of Jasher and Enoch, become known among the Mormons of Nauvoo.[36]

## Astronomical Pluralism in the Mormon Canon

Concerning the development of the plurality concept, intellectual historian Arthur Lovejoy has suggested five innovations implied in the new heliocentric cosmology: (1) other planets of our solar system are inhabited by living, sentient, and rational beings; (2) the closed world of medieval cosmology was replaced with an infinite universe;[37] (3) fixed stars are suns similar to our own and surrounded by planetary systems; (4) these planets are inhabited by conscious beings; and (5) an infinite number of solar systems exist.[38] As the plurality-of-worlds doctrine emerged in Jacksonian America, it reflected its old world roots, conforming in broad outlines to Lovejoy's scheme. Following the appearance of the books of Moses, Abraham, and the Doctrine and Covenants, Mormon writers later began to develop this theme more fully.[39]

As early as 1830, however, Joseph Smith first presented his ideas on multiple world systems within the context of Old Testament studies, justifying his own brand of astronomical pluralism as part of a long tradition of religious speculations on the subject. It has been argued that Psalms 8:3–4 has historically served as the point of departure for treatises on the plurality of worlds:[40]

> When I consider thy heavens, the work of thy fingers, the moon and the
> stars, which thou hast ordained; What is man, that thou art mindful of him?
> and the son of man, that thou visiteth him?

Here the suggestion is made that human beings, as one of God's creations, are no more significant than the creations of God elsewhere—
on *other* planets! But in Joseph Smith's case both the Old and New
Testaments provided material in new and innovative ways.

In the process of revising these sacred books, Joseph Smith presented
new meanings of Genesis and sought for new understanding of celestial
cosmology. In June 1830 Joseph Smith received his 'Visions of Moses':

> And he beheld many lands; and each land was called earth, and there were
> inhabitants on the face thereof. And worlds without number have I created;
> and I also created them for mine own purpose; and by the Son I created
> them, which is mine Only Begotten. But only an account of this earth, and
> the inhabitants thereof, give I unto you. For behold, there are many worlds
> that have passed away by the word of my power. And there are many that
> now stand, and innumerable are they unto man; but all things are numbered
> unto me, for they are mine and I know them. And the Lord God spake
> unto Moses, saying: the heavens, they are many, and they cannot be num-
> bered unto man; but they are numbered unto me, for they are mine. And
> as one earth shall pass away, and the heavens thereof even so shall another
> come; and there is no end to my works, neither to my words. (Moses 1:29,
> 33, 35, 37–38)

And in December he recorded the 'Prophecy of Enoch':

> And were it possible that man could number the particles of the earth, yea,
> millions of earths like this, it would not be a beginning to the number of
> thy creations; . . . Behold, I am God; Man of Holiness is my name; Man of
> Counsel is my name; and Endless and Eternal is my name, also. Wherefore,
> I can stretch forth mine hands and hold all the creations which I have made;
> and mine eye can pierce them also, and among all the workmanship of
> mine hands there has not been so great wickedness as among thy brethren.
> (Moses 7:30, 35–36)

Although some of these sources were not publicly presented until 1843,
they, together with Joseph Smith's scriptural Copernicanism (see chapter 5), formed the essential features of the Mormon concept of the
plurality of worlds during the early years of the church.[41]

As a whole, this early pluralistic view supports Lovejoy's five-fold
scheme. The ideas of inhabited planets and of an infinite number of
planetary systems are directly expressed in these verses of Moses, while
the infinity of space and stars with inhabited planets are all implied.
In other words, the basic features of astronomical pluralism were ev-

ident in Joseph Smith's thinking by December 1830. Moreover, in addition to this conventional view of the plurality of worlds, Joseph Smith also stated the simultaneous existence of multiple world systems throughout time itself. Again in Moses, we read

> There are many worlds that have passed away. . . . And there are many that now stand. . . . And as one earth shall pass away, and the heavens thereof even so shall another come; and there is no end to my works, neither to my words. (Moses 1:35, 38)

The pluralism developed early in Moses was carried over into Joseph Smith's increasingly mature and sophisticated theology. In February 1832 he and Sidney Rigdon received "The Vision" and, as published in *The Evening and the Morning Star*, it asserts "that by him [Christ], and through him, and of him, the worlds are made, and were created, and the inhabitants thereof are begotten sons and daughters of God; . . . worlds without end" (D&C 76:24, 112).[42] Later in December Joseph Smith recorded the "Olive Leaf" (D&C 88), and in May of the year following he received Doctrine and Covenants 93, both of which record astronomical pluralism conforming to Lovejoy's scheme and to the notion of pluralism in time.[43]

The Book of Abraham presents a detailed cosmology, featuring not only the plurality of worlds but an astronomy within which pluralism is an integral part. Perhaps the central feature of Abrahamic astronomy is the concept of governing worlds—places that apparently delimit and control the bounds and dimensions of other worlds:

> Kolob is set nigh unto the throne of God, to govern all those planets which belong to the same order as that upon which thou standest. . . . And he [God] put his hands upon mine eyes, and I saw those things which his hands had made, which were many; and they multiplied before mine eyes, and I could not see the end thereof. (Abraham 3:9,12)

Again, the plurality doctrine embedded in Abraham conforms to Lovejoy's scheme, while the notion that planets, or systems of planets, are controlled by other planets is an idea with some similarity to the views of the eighteenth-century cosmologist J. H. Lambert.[44]

In summary, Joseph Smith's views on the plurality of worlds clearly show that he espoused a position in keeping with Lovejoy's but also representing additional possibilities: (6) worlds have passed away and others have and are being formed (Moses 1:35, 38); (7) worlds are governed in a hierarchical relationship (Abraham 3:8-9); (8) every system of worlds has its own laws and bounds (D&C 88:36-38); (9) Christ made and/or makes all worlds (D&C 76:24; 93:9-10); (10) different kinds of people inhabit different worlds (D&C 76:112); (11) the Earth has been the most wicked of all worlds (Moses 7:36); (12) resurrected beings also reside on worlds (D&C 88:36–38); and (13) worlds exist both in space and time (Moses 1:35, 38; D&C 88:36–38, 42–47; 93:9–10).

Concerning the idea that this world is the most wicked of God's creations, Joseph Smith later wrote in Nauvoo:

> And I heard a great voice bearing record from Heav'n,
> He's the Saviour, and only Begotten of God—
> By him, of him, and through him, the worlds were all made,
> Even all that careen in the heavens so broad.
> Whose inhabitants, too, from the first to the last,
> And sav'd by the very same Saviour of ours;
> And, of course, are begotten God's daughters and sons,
> By the very same truths, and the very same pow'rs.[45]

Thus, (14) Christ's redemption is universal. Not all of these views, however, were original with Joseph Smith. For instance, the making of worlds (9) can be derived from the New Testament books Hebrews and Colossians; the philosophers Fontenelle, Swedenborg, and Kant all believed that different worlds are inhabited by different people (10); and worlds inhabited by resurrected beings (12) was a belief common among transmigrationists, a view that was discussed in Mormon periodicals.

Although there was very little exegesis of pluralism, cosmology, and science generally during the early church when Joseph Smith was alive, the framework within which pluralism is presented in the Mormon scriptures complements the basic theological ideas Joseph Smith had been developing.[46] In Moses pluralism is developed within the context of the inhabitants of God's creations, particularly their unrighteous nature. In Abraham the focus shifts to an hierarchical ordering of premortal spiritual creation. Doctrine and Covenants 76 deals almost exclusively with postmortal conditions and the characteristics of the various Mormon heavens; the context of section 88 presents specific discussion of the laws governing these kingdoms. These latter two sections describe not only the several heavens of Mormon cosmology but also their conditions, binding laws, and inherent bounds. Hence, Joseph Smith's concept of multiple inhabited worlds is more properly seen as a cosmological pronouncement of religious and metaphysical import than as speculation to convince the unbeliever in the truth of Christianity.

## Pluralist Thought and Joseph Smith's Cosmology

The ready availability of the concept of a plurality of worlds on the American frontier in the 1820s is obvious. This is not to suggest, however, that accessibility to the idea constitutes sufficient evidence that Joseph Smith derived his notion of multiple inhabited world systems exclusively from his cultural environment. The mere availability of pluralist views is in itself not an adequate argument for Joseph Smith's *coherent* system of beliefs. Ideas by themselves do not form an integrated and consistent system without the dimensions of a broader conceptual structure. Here, that basis is to be found neither in natural theology nor in a response to deism, but is uniquely cosmological. Although it is not clear that Joseph Smith's ideas on the plurality of worlds were consciously developed as a coherent whole, it is certainly the case that they are cosmologically self-consistent.

Ultimately, the question is not whether Joseph Smith was acquainted with Chalmers's *Astronomical Discourses* or Dwight's *Theology*, for example. Reading the Book of Moses closely, one cannot fail to be impressed with the repeated reference to vast numbers of creations. Not only do we read of "millions of earths like this," but also "the heavens cannot be numbered" and "worlds without number." The overwhelming impression is one of awesome size and grandeur. Paine had earlier argued that such conditions imply the absurdity of the atonement. Whereas Chalmers suggested there was nothing contradictory or absurd in the claim that God could use the Earth and its inhabitants to work out the universal atonement, Dwight believed only the Earth was in need of redemption. Joseph Smith, on the other hand, while assuming Chalmers's assertion and implicitly denying Dwight's, provided (under one interpretation) perhaps the most innovative alternative:

> Wherefore, I can stretch forth mine hands and hold all the creations which I have made; and mine eye can pierce them also, and among all the workmanship of mine hands there has not been so great wickedness as among thy brethren. (Moses 7:36)

In the context of Enoch's discussion on the plurality of worlds, this verse justifies pluralism in light of the skeptics' most serious argument against Christianity.[47]

Besides agreement on the plurality of worlds, there are other similarities between Dick's *Philosophy* and the emerging theology of Joseph Smith. The two most prominent features deal with the throne of God and the perfectibility of man, both of which Brodie emphasizes. Although she implies that Joseph Smith derived his notion of Kolob from Dick's idea of the throne of God, Dick views God as omnipresent, universal, and ethereal, which would preclude Joseph Smith's idea of a universal center upon which God, as being, dwells. Joseph Smith, and many others, shared Dick's and the eighteenth-century *philosophes'* view of the perfectibility of man.[48] But Joseph Smith argued for the ultimate *divine* perfectibility of humans, a concept Dick rejected. On such crucial doctrines as the attributes of God and His place of dwelling, the concept of eternal progression, creation *ex nihilo*, and the eternal nature of matter, there is also wide divergence of belief. Moreover, Dick espoused a dualistic metaphysics, while Joseph Smith became a strict monist. Theologically, Dick claimed that humans are utterly contingent upon God, while Joseph Smith eventually argued that humans are necessary.[49] On the nature of evil, sin, and the fall, the two also held polar views. After an exhaustive analysis of external evidence and doctrinal issues dealing with God, man, salvation, and other theological and metaphysical views, at least one contemporary writer, Edward T. Jones, concluded there are so few similarities in their thinking that Brodie's assertion must be rejected.[50]

Are there, however, as Brodie asserts, external reasons to justify her claim that Joseph Smith had read Dick's *Philosophy* prior to producing the Book of Abraham? Although Jones has shown that Brodie made numerous incorrect conclusions in trying to identify Joseph Smith's possession of Dick's *Philosophy*, Oliver Cowdery knew the book, or excerpts. In the *Messenger and Advocate* for December 1836, Oliver as editor quoted from Dick's *Philosophy* on, among other things, the plurality-of-worlds doctrine. In a later issue Oliver's brother Warren further inserted quotes from Dick's book *The Philosophy of Religion; or, An Illustration of the Moral Laws of the Universe*, published in 1826.[51] The first part speculated on the moral relations and conditions of extraterrestrial intelligences, while the second dealt with the foundations of morality. What cannot be ascertained without additional evidence is whether Joseph Smith was acquainted early in his career with Dick's writings. Even if he were, it seems highly unlikely that Joseph Smith benefited significantly from Dick's ideas. Although later, during his Nauvoo years, Joseph Smith possessed a copy of Dick's *Philosopher*, which in January 1844 he donated along with about forty of his own books to the recently organized Nauvoo Library and Literary Institute, he had already extensively expounded upon the subject six years prior to the Dick references appearing in the *Messenger and Advocate*.[52]

The primary sources for astronomical pluralism during the first third of the nineteenth century, Paine, Chalmers, Dwight, and Dick, were all widely known among the American reading public, with Dwight best known to the religious community. In the definitive study of the development and diffusion of astronomical pluralism throughout this period, historian Michael J. Crowe has analyzed nearly every published source of the concept appearing in the English-speaking world.[53] The breadth of literature dealing with pluralism is astounding, and, in addition to the above, it may be grouped into the following categories: (1) that rejecting pluralism as irreconcilable with Christianity (e.g., Walpole); (2) that rejecting pluralism as absurd (e.g., Coleridge); (3) that accepting pluralism as opposed to deism and supporting Christianity (e.g., Fuller, Nares, Harrington, Mitchell, Chalmers); (4) that rejecting Chalmers's particular advocacy of pluralism (e.g., Fergus, Overton, Maxwell); (5) that accepting pluralism as rejecting religion (e.g., Shelley, Byron); (6) that advocating pluralism as reconcilable with religion (e.g., Dwight, Swedenborg); (7) that advocating pluralism as science (e.g., Dick, Herschel, Copland); and (8) variations of the above.

The majority of these books were of minor importance or derived their arguments from the works of Chalmers, Dick, or Paine. A major purpose of many of these writers was not to explain science to the public so much as to provide arguments for the Christian message. Their arguments almost always took the form of extrapolation to the sciences of the day, of analogy to the human habitation, and of a teleological approach to God and the universe. Joseph Smith's version of pluralism, however, does not fit any of these categories easily. It is true that he could have supported positions (3), (4), and (6). Doing so, however, would have considerably altered the purposes for which he constructed his cosmology and would have compromised the terms in which he developed his views. He assumed pluralism—without defense! Astronomical pluralism, in Joseph Smith's version, possessed its own unique foundation, which, in the final analysis, was based on the emerging theology of the Restoration. Finally, Joseph Smith's writings on the pluralist question were never based on contemporary science nor did he argue by analogy or use a teleological approach. Because his views were presented in a variety of sources spread across a decade, however, it is not clear whether he built his system deliberately or otherwise. But from whatever sources Joseph Smith derived his views on the plurality of worlds, he developed them into a coherent system different from available sources.

## Joseph Smith, Astronomical Pluralism, and Cosmology

In assessing the merits of the environmental thesis—the view that one's cultural matrix is entirely sufficient to account for the emergence of a coherent set of ideas or conventions—it has not been necessary to adopt the view that Joseph Smith as a religious leader must always be seen as either a prophet or a charlatan, a dichotomy that has prevented useful and productive understanding of an enigmatic character. It is perfectly consistent with Joseph Smith's own experiences and writings to see him as one who was as much a part of the process of religious innovation as its primary medium of expression.[54]

The idea that Joseph Smith may have borrowed from cultural sources cannot, of course, be totally discounted—or confirmed. Yet asserting indigenous sources requires at the very least an explanation of both his deviations from available sources and his integration of his pluralistic ideas with his scriptural writings. In the pluralist concept Joseph Smith seems to have deviated significantly from the mainstream of those writing on this subject, whether evangelical or deistic. Furthermore, in so doing Joseph integrated his ideas on astronomy into a cosmological framework of complex dimensions.

Although we have explored only one aspect of his thinking, his pluralism was primarily an excursion into metaphysics and cosmology, rather than into natural theology. Joseph Smith was not trying to substantiate the Christian faith by association with the prevailing concepts and theories of science and philosophy. Neither was he interested in debunking deism. He felt no special need to do so, because he personally, and the Smith family generally, did not feel threatened by deistic arguments. If he were, he would have emphasized pluralism as an example of natural theology. But none of this appears. Given the available data, it seems reasonable to conclude (1) the plurality-of-worlds idea thoroughly saturated both the scientific and religious communities of Joseph Smith's time; (2) among the religiously orthodox, pluralism was used almost entirely as an instance of natural theology to substantiate the Christian message as a bulwark against skepticism; (3) Joseph Smith himself probably first encountered the idea of the plurality of worlds within the oral traditions of his times; (4) he likely did not use available literary materials (Chalmers, Dwight, Paine, Dick, etc.) as primary sources for his own version of pluralism; and (5) although he would not have rejected the conclusions of his religious contemporaries on the idea of the plurality of worlds, his own version extended far beyond theirs into cosmological and eschatological issues.

NOTES

1. Hugh Nibley has treated the plurality of worlds in his "Strange Thing in the Land: The Return of the Book of Enoch, Part 11," *Ensign* 7 (April 1977), 78–89. For a highly insightful and important examination of pre-Christian cosmology, including notions of the plurality of worlds, see Nibley's essential essay "Treasures in the Heavens: Some Early Christian Insights into the Organizing of Worlds," *Dialogue* 8, nos. 3/4 (Autumn/Winter 1974), 76–98, re-

printed in H. Nibley, *Old Testament and Related Studies*, ed. J. W. Welch, G. P. Gillum, and D. E. Norton (Salt Lake City: Deseret, 1986), 171–214. Recently Michael J. Crowe has examined Joseph Smith and the emergence of the plurality concept in Mormonism in his "Extraterrestrials and Americans," *The Extraterrestrial Life Debate, 1750–1900: The Idea of a Plurality of Worlds from Kant to Lowell* (New York: Cambridge University Press, 1986).

2. Thomas Paine, *The Age of Reason*, Part I, ed. Alburey Castell (Indianapolis: Bobbs-Merrill, 1957), 47.

3. Marjorie H. Nicolson, "Thomas Paine, Edward Nares, and Mrs. Piozzi's Marginalia," *Huntington Library Bulletin* 10 (1936), 107–8; also see R. C. Roper, "Thomas Paine: Scientist-Religionist," *Scientific Monthly* 58 (1944), 101–11.

4. Paine, *Age of Reason*, 50.

5. Ibid., 49.

6. Merle Curti, *The Growth of American Thought*, 3d ed. (New York: Harper and Row, 1964), 153.

7. Nicolson, "Thomas Paine," 114. Also see Vernon Stauffer, *New England and the Bavarian Illuminati* (New York: Columbia University Press, 1919), 75–76.

8. Richard L. Anderson, *Joseph Smith's New England Heritage* (Salt Lake City: Deseret Book, 1971), 207.

9. Richard L. Bushman, *Joseph Smith and the Beginnings of Mormonism* (Urbana: University of Illinois Press, 1984), 38. Also see Marvin S. Hill, "Secular or Sectarian History? A Critique of *No Man Knows My History*," *Church History* 43 (1974), 90.

10. Marvin S. Hill, "The Shaping of the Mormon Mind in New England and New York," *Brigham Young University Studies* 9 (Spring 1969), 357–63. Joseph Smith's religious milieu, stretching from Vermont, the New York lake country, and much of Pennsylvania to the Western Reserve of Ohio, was saturated not so much with infidelity as with religious contention for authority. Joseph Smith prayed, asking not if Christianity was true (skepticism) but which religious sect had authority. See Timothy L. Smith, "The Book of Mormon in a Biblical Culture," *Journal of Mormon History* 7 (1980), 3–21. On the thesis that Joseph Smith had written the Book of Mormon in response to deism, see Robert N. Hullinger, *Mormon Answer to Skepticism: Why Joseph Smith Wrote the Book of Mormon* (St. Louis: Clayton, 1980).

11. Bushman, *Joseph Smith and the Beginnings of Mormonism*, 5–6.

12. See Richard L. Bushman, *Joseph Smith and Skepticism* (Provo: Brigham Young University Press, 1974); and William G. McLoughlin, *Revivals, Awakenings, and Reform: An Essay on Religion and Social Change in America, 1607–1977* (Chicago: University of Chicago Press, 1978), 99–105, 108–11.

13. Edward Hitchcock, "Introductory Notice to the American Edition," in [William Whewell], *The Plurality of Worlds* (Boston: Gould & Lincoln, 1854), x.

14. See Dorothy Ann Lipson, *Freemasonry in Federalist Connecticut* (Princeton: Princeton University Press, 1977), 81–83.

15. For a succinct analysis of the *Astronomical Discourses*, see D. Cairns, "Thomas Chalmers's Astronomical Discourses: A Study in Natural Theology," *Scottish Journal of Theology* 9 (1956), 410–21.

16. Chalmers, *Astronomical Discourses*, 54. The complete psalmist reference is Psalm 8:3–4, also found in ibid., 13. See John H. Brooke, "Natural Theology and the Plurality of Worlds: Observations on the Brewster-Whewell Debate," *Annals of Science* 34 (1977), 221–286 (259).

17. Chalmers, *Astronomical Discourses*, 54.

18. Ibid., 122.

19. Crowe, *The Extraterrestrial Life Debate*, 188.

20. Ibid., 175. Dwight's sermons were published in five volumes as *Theology Explained and Defended in a Series of Sermons* (Middletown, Conn.: Clark & Lyman, 1818).

21. Charles E. Cunningham, *Timothy Dwight, 1752–1817* (New York: Macmillan, 1942), 330; and Roland Bainton, *Yale and the Ministry* (New York: Harper, 1957), 77. To date, we only have relatively firm information on one minister, the Reverend George Lane, who may have had contact with Joseph Smith during the early 1820s. Lane was an itinerant Methodist preacher involved in the revivals associated with the reawakening of religious sensibilities during this period. Whether or not his sermons made reference to astronomical pluralism is not known, but being a neo-orthodox revivalist and believing that the spread of infidelity would undermine the Christian faith, he probably dwelt occasionally on this topic. See Larry C. Porter, "Reverend George Lane—Good 'Gifts,' Much 'Grace,' and Marked 'Usefulness,' " *Brigham Young University Studies* 9 (Spring 1969), 321–40. It is highly unlikely, however, that the pluralist doctrine would have been developed in the spoken word with the same degree of detail—or lack thereof—as represented in the writings of Dwight or Chalmers. Moreover, and perhaps of greater significance, the contexts within which Joseph Smith developed astronomical pluralism were simply inappropriate as topics needed to fan the fires of revivalistic enthusiasm.

22. See, for instance, sermons 5, 6, 7, 13, 17, 42, in Dwight, *Theology Explained*.

23. Ibid., 5:508.

24. Fawn M. Brodie, *No Man Knows My History: The Life of Joseph Smith, the Mormon Prophet* (New York: Alfred A. Knopf, 1946), 171. Also, Klaus Hansen has suggested a similar correspondence in his *Mormonism and the American Experience* (Chicago: University of Chicago Press, 1981), see 79–80.

25. For a complete analysis of this issue, see my "Joseph Smith and the Manchester (New York) Library," *Brigham Young University Studies* 22, no. 3 (Summer 1982), 333–56.

26. For a discussion of these sources, see ibid., and Milton V. Backman, Jr., *Joseph Smith's First Vision*, 2d ed. (Salt Lake City: Bookcraft, 1980), 47–52.

27. For a complete analysis of this issue, see my "Joseph Smith and the Plurality of Worlds Idea," *Dialogue* 19, no. 2 (Summer 1986), 13–36, particularly 21–25.

28. The first weekly newspaper in Palmyra was the *Palmyra Register* (1817–21), followed by the *Western Farmer* (1821–22) and the *Palmyra Herald, Canal Advertiser* (1822–23), all published by Timothy C. Strong. In 1823 Strong sold his paper to Pomeroy Tucker and E. B. Grandin, who superseded Strong's paper with the *Wayne Sentinel*. In 1828 the *Palmyra Freeman* (1828–29) began publishing, as did the *Reflector* (1829–30), a short-lived serial perhaps best known for its illicit printing of portions of a pirated copy of the Book of Mormon manuscript. Prior to the formation of Wayne County in 1823, the county seat for Palmyra was Canandaigua in Ontario County. The town of Canandaigua, about eight miles south of the Smith farm, published a variety of newspapers at the time, including the *Ontario Messenger*, the *Ontario Repository*, the *Ontario Republican*, and the *Ontario Freeman*.

29. *Wayne Sentinel*, 12 May, 14 July 1824, 1 December 1826.

30. *Palmyra Register*, 28 July, 11 August, 15 December 1819, 22 March 1820; *Wayne Sentinel*, 6, 27 March, 9, 16, 23 October 1829. On deism, see *Palmyra Repository*, 10 March 1818, 7 February 1821. On science generally, see "From Chalmers' Sermons," *Ontario Repository*, 25 May 1825; "From Dick's Christian Philosopher," *Wayne Sentinel*, 23 March 1827; and "General Spectacle of the Universe," ibid., 22 August 1828; and "Varieties of Nature," *Palmyra Register*, 24 November 1819.

31. Marion B. Stowell, *Early American Almanacs: The Colonial Weekly Bible* (New York: Burt Franklin, 1977), ix.

32. Ibid., xiv–xvii.

33. Ibid., 164–66, and passim.

34. Since the early 1840s it has been frequently suggested that Joseph and Hyrum Smith, and Brigham Young and others close to the Mormon prophet, plagiarized Masonic mysticism as a ruse to embellish the Mormon concept of the temple with a form of legitimacy. See Reed C. Durham's 1974 Mormon History Association presidential address entitled "Is There No Help for the Widow's Son?" in *An Underground Presidential Address*, ed. Mervin B. Hogan (n.p., n.d.). Durham's address is liberally sprinkled with unsubstantiated claims, however, and is far more suggestive than either definitive or substantive. Perhaps the most extensive examination of Mormon and Freemasonry connections has been undertaken by Mervin B. Hogan. Among his many related publications are *The Origin and Growth of Utah Masonry and Its Conflict with Mormonism* (Salt Lake City: Campus Graphics, 1978); *Mormonism and Freemasonry under Covert Masonic Influences* (Salt Lake City: n.p., 1978); *Mormonism and Freemasonry: The Illinois Episode* (Salt Lake City: Campus Graphics, 1980); and *Freemasonry and the Lynching at Carthage Jail* (Salt Lake City: n.p., 1981). For additional themes dealing with Mormonism and Freemasonry, see C. Mark Hamilton, "The Salt Lake Temple: An Architectural Monograph" (Ph.D. dissertation, Ohio State University, 1978), 6–8. On Mormon temple symbolism, see idem, "The Salt Lake Temple: A Symbolic Statement of Mormon Doctrine," in *The Mormon People, Their Character and Traditions*, ed. Thomas G. Alexander (Provo: Charles Redd Monographs, 1980), 103–27; and on Mormon symbolism generally, Allen D. Roberts, "Where Are the All-Seeing

Eyes? The Origin, Use and Decline of Early Mormon Symbolism," *Sunstone* 4 (May/June 1979), 22–29. For an invaluable, and mammoth, exegesis of Masonic ceremony and mysticism, see Albert Pike, *Morals and Dogma of the Ancient and Accepted Scottish Rite of Freemasonry* (Richmond, Va., 1871), chap. 28, particularly 595–98. The 1924 edition contains a very valuable index (218 pages) to Pike's *Morals* by T. W. Hugo.

35. Though containing the most representative collection of classical Greek and Latin works, the library of Thomas Jefferson, perhaps the finest library in America, contained only one book by Proclus—*Philosophical and Mathematical Commentaries on the First Book of Euclid's Elements* (London, 1792). The more relevant *Six Books of Proclus on the Theology of Plato* (1816), trans. Thomas Taylor, was virtually unknown at the time. For possible Pythagorean connections with Abrahamic astronomy, see William Dibble's suggestive essay "The Book of Abraham and Pythagorean Astronomy," *Dialogue* 8, nos. 3/4 (Autumn/Winter 1973), 134–38.

36. "The Book of Jasher," *Times and Seasons* 1 (June 1840), 127; and "The Apocryphal Book of Enoch," *Millennial Star* 1 (July 1840), 61–63. For a recent examination of the importance of the Book of Enoch in the development of Mormon theology, particularly regarding the concept of Zion, see the provocative essay by Steven L. Olsen, "Zion: The Structure of a Theological Revolution," *Sunstone* 6 (November/December 1981), 21–26. Recently, Hugh Nibley has noted that the "Ethiopic" manuscript of Enoch was available in translation by 1821 and that much of the scholarly translation is remarkably similar in style and meaning to Joseph Smith's much shorter version (Book of Moses). To explain this alleged coincidence, Nibley has argued that both Enoch and Joseph Smith received their knowledge from the same (divine) source. The veracity of Nibley's argument hinges on two essential points: (1) whether Joseph Smith had in his possession a clandestine copy of the translated Enoch sometime prior to 1829–30; and (2) the similarity of style and substance between the translated Enoch and Joseph Smith's Enoch. Nibley's full examination of this issue appeared as "A Strange Thing in the Land: The Return of the Book of Enoch," published in thirteen parts in *Ensign* from October 1975 to August 1977; reprinted in H. Nibley, *Enoch the Prophet*, ed. S. D. Ricks (Salt Lake City: Deseret Book, 1986), 91–305 (particularly 236–40). For a recent translation of the "astronomical" sections of Enoch, see O. Neugebauer, "The 'Astronomical' Chapters of the Ethiopic Book of Enoch (72–82)," *K. Danske Matematisk-fysishe Meddelelser* [Copenhagen] 4, no. 10 (1981), 1–42.

37. See Alexandre Koyre, *From the Closed World to the Infinite Universe* (Baltimore: Johns Hopkins University Press, 1957), passim.

38. Arthur O. Lovejoy, *The Great Chain of Being: A Study of the History of an Idea* (Cambridge, Mass.: Harvard University Press, 1936), 108.

39. Although writers within the church referred to the idea of celestial pluralism as early as 1832, the most significant, noncanonized development occurred first during the Nauvoo years of Joseph Smith, and later in Utah by leading Mormon authorities (see chap. 5).

40. See, for instance, Brooke, "Natural Theology and the Plurality of Worlds," 259. In his inspired translation of the Bible, Joseph Smith did not alter the text of these verses in Psalms.

41. Moses 1 was received in June 1830 (Colesville, New York) and first published in the *Times and Seasons* of January 1843 (71–73), while Moses 7 was received in mid-December of 1830 (Fayette, New York) and published in August 1832 in *The Evening and the Morning Star*, 2–3. For a discussion of the dates when these revelations were received, see Robert J. Matthews, *"A Plainer Translation": Joseph Smith's Translation of the Bible, a History and Commentary* (Provo: Brigham Young University Press, 1975), chap. 11, particularly 72, 221–24.

42. I have retained the phraseology of this revelation as it was originally published in *The Evening and the Morning Star* 1 (July 1832), [10–11]. Received at Hiram, Ohio, D&C 76 was later printed in the *Millennial Star* 2 (May 1841), 17–21, and in the *Times and Seasons* 5 (1 August 1844), 592–95.

43. See D&C 88:36–38, 42–47; D&C 93:9–10. Both revelations were given at Kirtland, Ohio. For the historical background of these revelations, including their dates, see Lyndon W. Cook, *The Revelations of the Prophet Joseph Smith: A Historical and Biographical Commentary of the Doctrine and Covenants* (Provo, Utah: Seventy's Bookstore, 1981), passim.

44. Johann H. Lambert, *Cosmological Letters on the Arrangement of the World-Edifice*, trans. with introduction and notes by S. L. Jaki (New York: Science History Publications, 1976), passim. Also see Michael A. Hoskin, "The Cosmology of J.H. Lambert," *Stellar Astronomy: Historical Studies* (Bucks, England: Science History Publications, Ltd., 1982), 117–23. The idea of hierarchical control had been suggested in 1832; see D&C 88:42–44.

45. Joseph Smith, *Times and Seasons* 4 (1 February 1843), 82–85.

46. For a discussion of the emergence of a Mormon theology, see Thomas Alexander, "The Reconstruction of Mormon Doctrine: From Joseph Smith to Progressive Theology," *Sunstone* 5, no. 4 (1980), 24–33; T. Edgar Lyon, "Doctrinal Development of the Church during the Nauvoo Sojourn, 1839–1846," *Brigham Young University Studies* 15, no. 4 (1975), 435–46; and Van Hale, "The Doctrinal Impact of the King Follett Discourse," *Brigham Young University Studies* 18, no. 2 (1978), 209–25. Also see, most recently, Grant Underwood, "Book of Mormon Usage in Early LDS Theology," *Dialogue* 17, no. 3 (1984), 35–74.

47. Not all persons who actively engaged in natural theology to substantiate their religious (Christian) faith and who espoused the pluralist doctrine felt threatened by Paine and other skeptics. See, for instance, Henry Fergus, *An Examination of Some of the Astronomical and Theological Opinions of Dr. Chalmers* (Edinburgh, 1818).

48. See Carl Becker, *The Heavenly City of the Eighteenth-Century Philosophers* (New Haven: Yale University Press, 1932).

49. See Blake Ostler, "The Idea of Pre-Existence in the Development of Mormon Thought," *Dialogue* 15, no. 1 (1982), 59–78.

50. Edward T. Jones, "The Theology of Thomas Dick and Its Possible Relationship to That of Joseph Smith" (M.A. thesis, Brigham Young University, 1969), passim.

51. *Messenger and Advocate* 3 (December 1836), 423–25; ibid., 3 (February/March 1837), 461–63, 468–69.

52. Kenneth W. Godfrey, "A Note on the Nauvoo Library and Literary Institute," *Brigham Young University Studies* 14 (1974), 386–89.

53. See Crowe, *The Extraterrestrial Life Debate,* passim. In addition to the published sources discussed here, the pluralist works during this period (1790–1830) exerting peripheral influence include A. Fuller, *The Gospel Its Own Witness; or, The Holy Nature and Divine Harmony of the Christian Religion Contrasted with the Immorality and Absurdity of Deism* (1799–1800); E. Nares, *An Attempt to Shew How Far the Philosophical Notion of a Plurality of Worlds Is Consistent, or Not So, with the Language of the Holy Scriptures* (1801); R. Harrington, *A New System on Fire and Planetary Life* (1796); J. Mitchell, *On the Plurality of Worlds* (1813); Anon., *A Free Critique of Dr. Chalmers's Discourses on Astronomy* (1817); H. Fergus, *An Examination of Some of the Astronomical and Theological Opinions of Dr. Chalmers* (1818); J. Overton, *Strictures on Dr. Chalmers' Discourses on Astronomy* (1817); A. Maxwell, *Plurality of Worlds* (1817); A. Copland, *The Existence of Other Worlds* (1834).

54. See John F. C. Harrison, *The Second Coming: Popular Millenarianism, 1780–1850* (New Brunswick: Rutgers University Press, 1979), 176–92; W. H. Oliver, *Prophets and Millennialists: The Uses of Biblical Prophecy in England from the 1790s to the 1840s* (London: Oxford University Press, 1978), 218–38; Gordon S. Wood, "Evangelical America and Early Mormonism," *New York History* (October 1980), 359–86; Paul E. Johnson, *A Shopkeeper's Millennialism: Society and Revivals in Rochester, New York, 1815–1837* (New York: Hill and Wang, 1978); and, most recently, Grant Underwood, "Millenarianism and the Early Mormon Mind," *Journal of Mormon History* 9 (1982), 41–51. For an extremely insightful view of Joseph Smith as such a complex individual, see Jan Shipps, "The Prophet Puzzle: Suggestions Leading toward a More Comprehensive Interpretation of Joseph Smith," *Journal of Mormon History* 1 (1974), 3–20; and most recently, Bushman, *Joseph Smith and the Beginnings of Mormonism.* Earlier, Bushman explored the environmental thesis on the issue of political themes in his "Book of Mormon and the American Revolution," *Brigham Young University Studies* 17, no. 1 (1976), 3–20, reprinted in *Book of Mormon Authorship: New Light on Ancient Origins,* ed. Noel B. Reynolds (Provo: Religious Studies Center, 1982), 189–211.

# 5

## Cosmology and Mormon Thought

While Christ and the atonement remain the central feature of the Mormon religious message, for Mormons the atonement became understandable most forcefully in the context of a universal vision that encompasses past, present, and future states of humankind. Questions dealing with themes such as where humans have come from, why they are here, and where they are going have remained central. Consequently, it is simply not possible to overestimate the degree to which Mormons came to regard—implicitly, if not otherwise—the plurality of worlds as a basic conceptual category needed to integrate their entire beliefs. To misunderstand Mormon pluralism is virtually to deny the essence of the Mormon cosmic vision.

Mormon theology suggests an enormously complex, yet relatively coherent, worldview. During the 1830s Joseph Smith laid much of the scriptural foundation for these developments by exploring four astronomical themes: (1) the heliocentric cosmology of modern science, (2) an (Abrahamic) cosmology that allowed for the direct interposition of God into His creation, (3) creation accounts of the world, and (4) the plurality-of-worlds doctrine. Joseph Smith integrated these four themes because all were needed to various degrees in order to develop a completely integrated theologic cosmology. Together, these ideas fundamentally support the pluralistic notion that multiple world systems are occupied by sentient beings of various degrees of perfection from premortal beings to several different penultimate worlds of resurrected glory. Moreover, other forms of pluralism—the plurality of wives and the plurality of gods—all fitted into Joseph Smith's emerging view of astronomical pluralism. He eventually even argued for a pluralist interpretation of the nature of being itself, suggesting that ontologically humans are a necessary (as opposed to a contingent) partner with God in the cosmology of the universe.

The theological speculations by the followers of Joseph Smith, particularly in the years following the Nauvoo exodus, assumed with increasing confidence the very basic idea of multiple worlds. Themes dealing with the relationship of humans to God, of the hierarchy of gods in an increasingly pluralistic universe, of the postmortal existence of resurrected beings, of the vertical relationships of the very creation itself, both physical and spiritual—all these and many more themes subsumed astronomical pluralism. These possibilities were not lost on the Mormon imagination, and as a result Mormon theology became saturated with this fundamental organizing idea. To accept the central feature of pluralism within Mormonism, however, is also to thrust pluralism directly into questions of science and religion.

The Mormon canon argues that both biblical and latter-day prophets and seers, such as Enoch, Abraham, Moses, and Joseph Smith among others, were shown visions of the heavenly realms to orient them to God's dominion and eternal purposes. In these manifestations, features of the celestial region suggest a total cosmic vision that provides the substructure for the entire theology of Mormonism. That structure offers a governing system of worlds and all stellar objects (Abrahamic astronomy) in which the total expanse of the heavenly regions is explained spatially (the plurality of worlds) and temporally (creation of the universe itself). Although not as central to Mormon theology as to other religions, heliocentric cosmology brings biblical and Mormon scriptural sources into harmony with contemporary science.

Whereas Joseph Smith had no pressing interest in harmonizing science with religion (indeed he had no significant interest directly in science itself), his vision of the full sweep of existence—premortal, mortal, and postmortal—urged him to construct a cosmology in which cosmogonic and eschatological issues prevailed. Ultimately, astronomical pluralism, Abrahamic astronomy, and creation accounts provided the spiritual and physical grounding for his nascent theology. But in so doing, his theologic cosmology propelled Mormonism ultimately into scientific issues.

## Mormon Heliocentrism

With the revelations of Joseph Smith and the writings of Enoch and Abraham, by 1836 the plurality of worlds had achieved a secure foundation in the emerging theology of Mormonism. As early as 1830

Joseph Smith had already endorsed the notion of heliocentrism (Copernicanism). In the Book of Mormon, mostly translated between early April and July of 1829, Copernicanism is presented explicitly. Thus, "it is the earth that moveth and not the sun" (Helaman 12:15), and "even the earth . . . and its motion . . . move in [its] regular form" (Alma 30:44).

Because a number of Old Testament passages historically have suggested a geocentric cosmology—principally those found in Joshua (10:12–14), Job (9:6–7), Isaiah (38:7–8), and II Kings (20:8–11)—in terms of biblical cosmology the Book of Mormon references appear ostensibly as a historical anachronism.[1] Using Joshua's command "Sun, stand thou still . . . and the sun stood still" as the strongest injunction implying the geocentric view, the obvious conflict between the traditional biblical exegesis of these Old Testament verses and Book of Mormon heliocentrism can be understood in one of several ways: either Old Testament geocentrism (taken literally) is incomplete (though not "wrong"), and therefore its modern interpretation is presumably faulty; or Old Testament references must be interpreted metaphorically and allegorically; or Joseph Smith's insertion of a stationary sun had been acquired largely from indigenous cultural sources; or the Book of Mormon as a historical document reflects an unorthodox cosmology of pre-Christian origins.

The entire Book of Mormon reference in Helaman suggests (at least astronomically) a stationary sun can be explained by shifting the point of reference from a moving sun (commanded to be still, as in Joshua) to a moving earth that when commanded to "goeth back . . . it appeareth unto man that [only] the sun standeth still" (Helaman 12:15). The argument given in Helaman is, by modern understanding, of course, correct: without additional observational data (discovered conclusively in 1838–39), it is not possible to distinguish (definitively) whether the earth or the sun (or both) is in motion.[2] Therefore, in one sense the Helaman account does not so much contradict Old Testament geocentrism as it simply stands these views upside-down, so to speak, in favor of the modern view of heliocentrism by emphasizing the *relative* position of the observer. If the Joshua reference, say, is understood metaphorically, the "correction" in Helaman does little to alter the significance of Joshua. Although this sort of allegorical interpretation has a significant non-Mormon tradition, it can also be sustained within a Book of Mormon exegesis. Immediately following the heliocentric reference in the book of Helaman, one reads: "For ye shall know of the rising of the sun and also of its setting" (Helaman 14:4). Because this reference cannot deny the immediately preceding, explicitly heliocentric scripture, one must interpret the "rising of the sun" allegorically. Consequently, rather than incompatible scriptural traditions, this sort of Book of Mormon exegesis of a heliocentric cosmology may be used to reinterpret Old Testament geocentrism.

A more serious criticism, of course, deals with the cosmological implications of a heliocentric view (as in Helaman) for an indigenous Old Testament cultural worldview (as in Joshua). With few exceptions, pre-Copernican views of the cosmos, whether Hebrew, Greek, or Egyptian, strongly favored a geocentric cosmology. If the Book of Mormon reflects Old Testament understanding of the cosmos, then the discrepancy can only be rationalized as being highly idiosyncratic. Nevertheless, though heliocentrism increasingly dominated Western thinking since the late Renaissance, a sun-centered worldview was not unknown in pre-Christian days.[3]

Whatever the source of these ideas, it is certainly true that Book of Mormon heliocentrism is entirely consistent with the post-Copernican astronomy of the nineteenth century. Although some of Joseph Smith's Mormon contemporaries argued that it would be possible for God to reverse the diurnal and annual motions of the planets, it was not probable because God cannot violate the laws of nature.[4] In this view of the cosmos, natural law is fundamentally dominant; only under the most extreme situations (such as New Testament miracles) are divine or magical forces permitted.

## Abrahamic Astronomy and the 'Universal Vision'

In the Mormon tradition the phrase "Abrahamic astronomy" has historically referred to some astronomical ideas presented in the Book of Abraham.[5] Perhaps the central feature of Abrahamic astronomy is the concept of governing worlds (see chapter 4). Briefly, the Book of Abraham suggests that God's physical dominion (throne) is located near a star called Kolob (Abraham 3:2). While it may seem reasonable to suppose that this throne would be located at the center of God's dominion and, therefore, that his dominion would coincide with some distinguishing feature of the universe, all efforts to identify it with, say, the center of the Milky Way Galaxy, the local cluster of galaxies, or the center of the universe itself are speculative. Wherever Kolob is located, however, its purpose is to "govern" all planets that are of the same "order" as the earth (Abraham 3:9).

Because the Abrahamic account does not sufficiently clarify these terms, it is not known whether this reference is to be understood primarily in physical or spiritual terms or in metaphorical and allegorical terms.[6] Thus, "to govern" might mean a physical bonding as with gravity, while "order" could conceivably mean planets similar to the earth in size, or planets in the same region of our galaxy or even in the entire Milky Way Galaxy, or even a number of other possibilities. Kolob also provides the light for all stars, including that for our sun (Abraham: Facsimile no. 2). Because modern astrophysics has shown that nuclear processes deep within a star's interior generate stellar heat and light, this reference can only mean spiritual light. Finally, Kolob controls the "set times" of the revolutions of various stellar bodies (Abraham 3). Physically, this reference conceivably refers to the axial, orbital, and differential revolutions of stellar objects.[7] If we interpret this in physical terms, however, perhaps we should also interpret the "light" statement in a physical sense.

Because many possibilities have been given historically, none of which are authoritative, it would seem best simply to note that Mormonism possesses great speculative powers as well as a tremendous penchant for things "scientific."

## Cosmogony and Mormon Creation Narratives

Two interrelated traditions in Mormonism, both introduced by Joseph Smith, deal with creation issues: the reinterpretation of a variety of cosmogonical ideas and the introduction of new canonical creation narratives. In the process of reflecting on the biblical creation accounts, Joseph Smith began using the term *creation* to mean that, rather than emerging ex nihilo, the earth was organized from "pre-existing materials." Moreover, as Joseph Smith argued, "the elements are eternal. That which has a beginning will surely have an end. . . . Every principle proceeding from God is eternal. . . . In the translation 'without form and void' should read 'empty and desolate.' The word created should be 'formed' or 'organized.' "[8] An ex nihilo creation process was therefore an impossibility.

Not only was the creation process as Joseph Smith understood it fundamentally opposed to the views of orthodox Christianity; his views reflected certain scientific ideas that later emerged over the course of the nineteenth century. About a decade after Joseph Smith reinterpreted the creation process, four European scientists simultaneously and independently discovered the first law of thermodynamics.[9] Known as the conservation of energy, this fundamental law of nature demands that although energy may change its state it can never be created or lost. Matter, as one form of usable energy, cannot be created out of nothing. Thus it was argued that Joseph Smith's view on creation was vindicated loosely by science, and only natural laws and processes—not supernatural interventions—were needed to organize and form the world. In a word, Mormons argued, orthodox Christianity a la creation ex nihilo had been corrupted by false principles.

To augment the creation account in Genesis, Joseph Smith introduced, in Moses and in Abraham, two further creation stories that have yielded several astronomically related images. Enoch (in Moses 2 and 3) outlines a spiritual creation of the heavens and the earth that preceded the physical creation of the world discussed by Abraham (in Abraham 4 and 5). Thus Enoch explains that God "created all things spiritually, before they were naturally upon the face of the earth" (Moses 3:5), asserting the spiritual nature of the cosmos. In this view God created all things—including presumably his offspring as well as the cosmos itself (or, in the latter case, at least the natural laws)— spiritually first and only afterwards patterned His creation physically.[10] Mormonism has historically viewed the creation account cosmogonically, as describing the origins of the world. Thus, Enoch uses the term *day* to distinguish the various creation periods; Abraham uses the more ambiguous word *time*, suggesting the complexity of the physical creation process itself.

Within Mormonism there exists an enormous literature devoted to the idea of the creation and its derivatives. In addition to the denial of creation ex nihilo and the affirmation of the creation as being twofold—first spiritual and then physical—Mormonism views the creation of all things as describing an *organizing principle*, as being good and not in vain, and as a preparation ultimately for humankind itself. Although occasionally asserting a very old earth (see chapter 8), the creation process is most frequently described as having occurred in one "week" of six "days" duration.

## Astronomical Pluralism in Early Mormonism

In their thinking about the plurality of worlds, and science generally, Mormons have used approaches familiar in other religious traditions. Most Mormons heavily employed the *theological* argument to support their views by referencing various scriptural sources, including Psalms 8:3–4, which traditionally had been used as the point of departure for speculations by Protestants in defense of the pluralistic view. Many invoked some form of classical *teleological* arguments which assume or argue a divine purpose in nature based on evidences of purposeful adaptation to ends. For example, God would not create a universe with countless stars if the stars did not have meaning beyond the simple fact of their existence. Hence, the universe was created for the purposes of life, and stars are the physical sources of life-bearing conditions needed to support intelligent life on planetary systems throughout the universe. One common form of the teleological argument, variously called the *cosmological* argument or argument by design, points to the causal order in the world and suggests that both common and scientific experience provides evidence of a purposive creator. In short, the universe was created with a plan (or design); it was not created in a haphazard or random way. Because there obviously is a design, goes the argument, there must have been (and still is!) a designer (God). The principle of *plenitude* holds that the creator of the world (the universe) would not create anything in vain. Therefore, for example, just as our star (the sun) has a planetary system with life, other stars must have planetary systems populated with sentient beings. Similarly, *analogical* arguments hold that the earth, the solar system, and everything contained in them, including living things, are types—that, for instance, rational, sentient life must exist on other planetary systems revolving about the myriad of stars precisely because that is the type of life known to humans in this system of multiple planets. *Eschatological* arguments, used with considerable frequency by Mormons and closely related to teleological thinking, rest on Christian doctrines that justify the present by reference to the end of the history of the world. For example, when all God's creations are bound together in their final quintessential and resurrected state, the end purpose of things will require the existence of countless worlds.[11]

Most Mormons couched their arguments first in a theological context. The scriptures of the restoration held a powerful grip on those who wrote on the topic, and they generally supported their views on the plurality of worlds from a religious vis-à-vis scientific/intellectual basis. The cosmological and teleological arguments also held particularly strong sway. Not until John A. Widtsoe and B. H. Roberts, however, did Mormons begin to exploit other arguments in support of the plurality of worlds. With respect to our solar system, these arguments suggested that all the known celestial and planetary bodies—including the moon and the sun—were likely inhabited by sentient beings in various stages of development. Speculations on the actual inhabitation of various celestial bodies were also extended to all the world systems of the "starry firmament," to use a phrase common in the nineteenth century. These views were held among both Mormons and non-Mormons; religious, scientific, and everyday people generally adhered to them.

After Joseph Smith presented his ideas on the plurality of worlds in his Enoch writings and in the Doctrine and Covenants (particularly sections 76 and 88), speculations on astronomical pluralism began to appear in 1832 in *The Evening and the Morning Star,* a monthly equivalent to today's *Ensign* magazine, published by W. W. Phelps. The first article to publicly present the plurality of worlds in a strictly Mormon forum was a reprint from a non-Mormon source that assumed the principle of plenitude:

> There are other worlds around us to which probably our earth with all its grandeur is but as dust in the balance. The eye wanders off enraptured with its discoveries amidst the bright orbs of heaven. Infinity of space is before it. Unnumbered spheres are above, and below, and around us. And when the eye is tired of gazing, and when its spirit flying vision has reached its utmost goal, it calls to its aid the benefits of scientific discovery, and stretches out into still more distant space, and there enjoys the new pleasure of seeing other worlds and beholding other wonders.[12]

Space, as the article suggests, is infinite in extent and contains an uncountable number of stars stretching without limit where unnumbered worlds possess wonders and display the grandeur of the creator. The November 1832 issue made tacit use of the teleological argument and contained additional comments concerning astronomical pluralism that "this world and the worlds beyond neither the sun, nor the moon, nor the planets, nor the stars, have ceased for a moment from performing their daily labors."[13]

Although *The Evening and the Morning Star* did not present many speculations on the notion of multiple world systems, the idea of the plurality of worlds was assumed in many contributions and given some discussion in a variety of topics. For instance, in the December 1833 issue there was an extensive discussion of the forthcoming millennium in which the author assumed the theological and plenitude arguments:

> It is a pleasing thing to let the mind stretch away and contemplate the vast creations of the Almighty; to see the planets perform their regular revolutions, and observe their exact motions; to view the thousand suns giving light to myriads of globes, moving in their respective orbits, and revolving upon their several axis [sic], all inhabited by intelligent beings; to consider that they all are visited with the light of his countenance, according to the revelation of his own character; that he communicates from time to time his will to all his creatures, and that he could not be impartial, were he to give a part the privilege of attaining to perfection and glory, and leave the other in darkness and uncertainty; but that Word by which all things were made will bring all alike to stand before him, and yet the least of all his

creatures will not be overlooked, though at the assemblage of worlds, but all will be rewarded according to their works.[14]

As the first church serial publication, *The Evening and the Morning Star* also continued in the American tradition of setting abstract ideas, such as pluralism, to poetic verse.[15]

After Oliver and Warren Cowdery became editors of the church's *Messenger and Advocate*, following the demise of *The Evening and the Morning Star* in September 1834, important extracts from Thomas Dick's books *Philosophy of a Future State* and *The Philosophy of Religion* were reprinted. Dick's published extracts helped to continue discussion of the plurality of worlds before the Mormon audience. "There are reasonings sufficient," wrote Oliver Cowdery in December 1836 concerning the reprinted sections, "to commend it [Dick's writing] to the attention of the reader."[16]

Once the Saints had settled in Nauvoo in 1839 and reestablished communal normalcy, Joseph Smith and other Mormon leaders began, though uncritically, to use the pluralist idea as a conceptual support for many other doctrines in order to express the limitless expanse of the Creator. Thus, Joseph Smith explained, translated beings are "held in reserve to be ministering angels unto many planets"; the Creator's kingdoms "are designed to extend till they not only embrace this world, but every other planet that rolls in the blue vault of heaven"; and "the grand councilors sat at the head in yonder heavens and contemplated the creation of the worlds which were created at the time."[17] Joseph Smith's ideas on pluralism as expressed in the seventy-sixth section of the Doctrine and Covenants were set to poetic, and later lyrical, form by Eliza R. Snow. Other poems also soon expressed variations of the cosmic vision of worlds, stars, systems, and endless creations in the church publications *Times and Seasons* and *Millennial Star*.[18]

Although Joseph Smith first developed the pluralist idea as a central theme in the still emerging theology of Mormonism, Parley P. Pratt and his brother Orson became the most prolific in developing the pluralist idea in poetic and scientific metaphor. Parley, the first Mormon authority to write religious tracts reflecting restorationist themes and also Mormonism's first poet, used cosmological ideas widely. Thus, in his poetic speculation "Regeneration and Eternal Duration of Matter" Parley surmised the creation process: "When in the progress of the endless works of Deity, the full time had arrived for infinite wisdom to organize this sphere, and its attendant worlds, and to set them in motion in their order amid the vast machinery of the universe. . . ."[19] Parley also used these ideas in his eschatological speculations: "To form a conception of heaven, I would contemplate all the planets of the solar system, of which our earth is but a small one, purified of all contamination which may have tainted them, and the inhabitants of each . . . gathered together, each on their own planet or sphere . . . living in common friendship and employing all their faculties in the service of God."[20]

While Parley engaged in poetic speculation, Orson Pratt felt more comfortable in developing the new cosmological ideas philosophically and scientifically. In an article entitled "Angels," Orson calculated the number of inhabitants in the solar system at one thousand million million residing on some thirty worlds, which in 1845 included seven planets of the solar system and, by Orson's estimate, about twenty-five other celestial objects.[21] Assuming cosmological, analogical, and plenitistic reasoning, Orson was concerned about arguing for an omniscient god whose glory was without bound: "If we were still to extend our calculations beyond the limits of the solar system, and take into consideration the inhabitants of the innumerable systems of worlds existing in the vast immensity of space, . . . our limited capacities [would be] incapable of conceiving any rational idea of the immense unlimited number of beings."[22]

Aside from these few presentations, mostly in the Mormon press, and the scriptural writings of Mormonism found in Enoch, Abraham, and the Doctrine and Covenants, no serious examination of astronomical pluralism and Mormon theology was undertaken during the Kirtland and Nauvoo periods of church history. In a completely different context from these literary sources, however, it has been suggested that sometime around 1837 Joseph Smith claimed that the moon was inhabited by sentient beings about six feet tall who dress in a Quaker style and live upwards of one thousand years. This view was first publicly mentioned in 1892 by Oliver B. Huntington. Huntington based his view on two sources: a description of moonmen allegedly given by Joseph Smith to Philo Dibble, which Huntington later transcribed into his own personal journal in 1881, and Huntington's own patriarchal blessing, given to him not by Joseph Smith, Sr., as he alleged, but by Oliver's father, William. Thus Huntington's reminiscences are third-hand at best and separated by half a century from the alleged events.[23]

There is no direct evidence, however, that Joseph Smith ever claimed that the moon or sun is inhabited. Others, including his brother Hyrum and later Brigham Young, however, held the view that both the moon and the sun are inhabited with living beings.[24] As a matter of fact, it would have been entirely consistent with his times (at least up to about 1835) had Joseph Smith claimed that the moon (or the sun) was inhabited. In 1835 the *New York Sun*, whose circulation was rather modest, ran a series of six daily accounts alleging the discovery of lunar inhabitants. These accounts were written with perfect journalistic flair and license, and, not surprisingly, within days the *Sun* had achieved the largest circulation of any paper in the country. In fact, the series of articles, written by Richard Adams Locke, became so successful that they were later republished in pamphlet form, widely cited, and therefore well known. Despite the fact that Locke later admitted that he had fictionalized the entire series, the public by and large continued to believe the account as described.[25]

This episode is an example of belief holding primacy over conflicting views. People had been culturally conditioned to believe in moonmen and sentient beings on other worlds for so many years that evidence to the contrary was simply ignored. As we have seen, many Christians had come to believe in the idea of inhabited worlds; and the moon was a perfectly reasonable candidate for such a place, because, among other things, it was believed to possess an atmosphere not unlike the earth's. Indeed, speculations on lunar inhabitation and sentient beings had increasingly entered the literary climate of Western thinking during the preceding two centuries.[26] In Locke's defense, however, his articles were not intended as a hoax on a credulous public. Rather, they were examples of an emerging native satirical genre that was soon to find full flavor in the writings of Edgar Allen Poe and Samuel Clemens.[27]

Concerning the Joseph Smith–moonmen episode, if in fact it actually occurred, there is no conclusive evidence one way or the other. One contemporary writer, Samuel W. Taylor, has expressed the view in his historical novel *Nightfall at Nauvoo* that Joseph Smith might have used the idea to jest with his friends, as he often was known to have done. After discoursing on the nature of the moon and its inhabitants, in private conversation with Joseph Smith, Eliza R. Snow, one of Joseph Smith's plural wives, later asked him how he knew so much about the matter. He replied simply that "a prophet always had to have an answer to every silly question."[28] Taylor depicts this [historical-fictional] episode with Eliza Snow as having occurred in Nauvoo during the fall of 1842—nearly seven years after Locke recanted his hoax. Although the plurality of worlds was an essential aspect of restorationist thinking, specific views about the moon were at best highly speculative and possibly treated by Joseph Smith, though not necessarily others, as such.[29]

Besides Locke, others on the American frontier about this time were also experimenting with the idea of inhabited worlds. Sometime around 1840 a Solomon Spalding wrote a 412-page manuscript, entitled "Romance of the Celes," that portrays a fanciful adventure into other worlds than ours. This Spalding is the nephew of his better-known namesake, who is known among Mormons as the author of an unpublished essay that has been alleged to have formed the basis of the Book of Mormon. The purpose of his story, Spalding tells us, is "to entertain all readers; to reclaim lifeless backsliders; lessen the prejudices of infidels, puzzle theologians and feast the real Christian and philanthropist who believes in both the justice and mercy of an all seeing God." In his romantic adventure that begins in Detroit, Spalding imagines sentient life throughout the solar system, including the moon, but particularly on Jupiter and Mercury. While "Thomas Dick has given us his views of the power and majesty of God, . . . [Spalding] attempts to finish the picture by romance, in order to fill up a great blank that presents itself to almost every son and daughter of Adam."[30] The manuscript was never published, and there is no evidence that Spalding exerted more than local influence with his ideas.

### Astronomical Pluralism after 1850

Following the introduction of astronomical pluralism, the endorsement of heliocentrism, the revelations of Abrahamic astronomy, and the creation narratives by Joseph Smith during the 1830s, Mormons, particularly ecclesiastical authorities, began to explore the many dimensions of a plurality of worlds presenting numerous arguments to support their views. As a result, ideas and attitudes about science and the relationship of science to religion began to emerge with growing confidence in the Mormon messianic vision of reality. Mormons increasingly employed empirical evidence and the observations of science to support their views of a religio-scientific cosmology. Reasons for doing so were twofold: (1) Mormon theology had become committed philosophically to a realist position, thus supporting a positive conception of science (understood as "true" science), and (2) Mormons saw the scientific world as providing confirmation to the revealed world of the restoration ("true" religion).

Yet Mormons never engaged thoroughly and consistently in natural theology, the basic premise of which asserts not only the compatibility of science and religion but also the primacy of science over religious interpretation in those areas where conflict occurs. Although the philosophical arguments and scientific evidences used to affirm religious claims have their value as rational supports, within Mormonism they have never been used as conclusive proofs of such claims. Among Mormons, natural theology has always been epistemologically suspect. Because the existence and nature of God are known only by revelation, both the primacy of revelation and the revealed understanding of God are protected. Therefore, the final arbiter of any alleged conflict between science and religion remains unchallenged.

Although noncanonical Mormon publications from 1832 to the Nauvoo exodus in 1846 and 1847 occasionally and briefly examined the plurality of worlds, extensive use and discussion of astronomical pluralism did not emerge until after the Saints had moved to the Great Salt Lake Valley in the 1850s. Virtually all speculations that dealt with the plurality of worlds—or, to use the phrase most commonly used at the time, 'worlds without number'—appear in various conference addresses later found in church-related publications such as the *Journal of Discourses* and the *Millennial Star* and in a few additional works such as Parley P. Pratt's *Key to the Science of Theology* (1855). Beginning around 1900 and until roughly 1930, speculations on science, cosmology, and the plurality of worlds became infused with the ideas of the first academically trained Mormon scientists (and later general authorities) James E. Talmage, John A. Widtsoe, and Joseph F. Merrill. With the possible exception of the Pratt brothers, however, it was B. H. Roberts, writing in the first several decades of the twentieth century, who contributed more than any other Mormon authority after the initial years of the restoration to the topic of astronomical pluralism.

Aside from the idea of worlds without number first proposed in the scriptures of the restoration and later developed in the sermons of Joseph Smith, the earliest, and most extensive, coherent presentation of astronomical pluralism among Mormon thinkers was given in the very popular book *Key to the Science of Theology*, written by Parley P. Pratt in 1855. *Science of Theology* was intended as a kind of prolegomena to the study of theology, "an introductory key," to use Pratt's words, "to some of the first principles of . . . divine science." Pratt advocated a positive conception of science, noting that "while every science, every art is being developed; while the mind is awakened to new thought; while the windows of heaven are opened, as it were, and the profound depths of human intellect are stirred, . . . religious knowledge seems at a standstill."[31]

In chapters 6 ("Origin of the Universe") and 16 ("Man's Physical and Intellectual Progress") Pratt argues for a plurality of worlds with certain spatial, temporal, and material conditions: (1) space has always existed as boundless and infinite; (2) matter or "elements" are coextensive with space; (3) elements are eternal, uncreated, and self-existing; (4) elements possess certain inherent attributes (such as being active, animate, psychic, and possessing extension); and (5) therefore elements are both physical and spiritual. Given a plurality of gods, whose spiritual offspring increase in numbers, both spiritual and physical worlds are needed in order to transplant them. In turn resurrected and eternal states of these beings would be required for their eternal progress, thus necessitating the "regeneration of worlds." As a result, suggested Pratt, "the heavens will multiply and new worlds and more people will be added to the kingdoms of the Fathers." Therefore, "unnumbered millions of worlds [planets and solar systems?] and of systems of worlds [galaxies?] will necessarily be called into requisition and be filled by man . . . and all the vast varieties of beings and things that ever budded and blossomed in Eden or thronged the hills and valleys of the celestial paradise." Eventually, "planets will be visited, messages communicated, acquaintances and friendships formed, and the sciences vastly extended and cultivated." All of this occurs beginning from the "great, central governing planet, or sun, called Kolob, until they are increased without number and widely dispersed and transplanted from one planet to another." In so doing, geography, astronomy, and "prophetic" science will be extended to millions of worlds.[32]

Pratt's views on astronomical pluralism also subsume Joseph Smith's ideas noted above and developed later by other Mormon authorities. Additionally, Pratt systematizes his views into a reasonably coherent whole. In order to support his cosmology, Pratt assumed both the teleological and the cosmological arguments that the universe is dominated by purpose and design. His ideas were also formulated within a theological context in order to support his version of eschatological issues. Since Pratt's *Science of Theology* was published shortly after the Nauvoo sojourn to the Salt Lake Valley and because of its coherence with the major themes emerging in Mormon theology, Pratt's book continued to influence directly and otherwise the views on astronomical pluralism within the Mormon hierarchy throughout the nineteenth century. Indeed, as evidence of its lasting effect on Mormon thinking, the *Science of Theology* has been reprinted eleven times from 1863 to 1979.[33]

Not only did Mormon authorities during the latter half of the nineteenth century accept the views on cosmology developed by Joseph Smith and Parley P. Pratt, but they also continued to expand and refine the notion of worlds without number. During the post-Nauvoo years, pluralist ideas were occasionally presented in publications such as the *Millennial Star* and various Mormon-published almanacs, particularly those by Orson Pratt and W. W. Phelps.[34] The most extensive use and development of astronomical pluralism occurred, however, mostly in the formal sermons of Mormon leaders given during general conferences of the church. As a result, virtually without exception, references

A remarkably detailed portrayal of the inhabitants of the Moon, from an English pamphlet published in 1836.

The 1835 moon hoax inspired this representation of lunar inhabitants and sundry creatures published in an English pamphlet in 1836.

# KEY

TO THE

# SCIENCE OF THEOLOGY:

DESIGNED AS

## An Introduction

TO THE FIRST PRINCIPLES OF SPIRITUAL PHILOSOPHY; RELIGION;
LAW AND GOVERNMENT; AS DELIVERED BY THE ANCIENTS,
AND AS RESTORED IN THIS AGE, FOR THE FINAL DEVELOPMENT
OF UNIVERSAL PEACE, TRUTH AND KNOWLEDGE.

### BY PARLEY P. PRATT.

O Truth divine! what treasures unrevealed,
In thine exhaustless fountains are concealed!
Words multiplied; how powerless to tell,
The infinitude with which our bosoms swell.

Liverpool:
F. D. RICHARDS, 15, WILTON STREET,
London:
L.D SAINTS' BOOK DEPOT, 35, JEWIN ST., CITY,
AND ALL BOOKSELLERS.

1855.

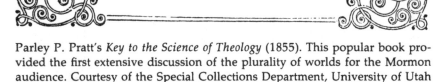

Parley P. Pratt's *Key to the Science of Theology* (1855). This popular book provided the first extensive discussion of the plurality of worlds for the Mormon audience. Courtesy of the Special Collections Department, University of Utah Libraries

to worlds without number occurred in the context of additional *theological* speculations. Typically, Mormon authorities would consider a particular religious or theological idea and incorporate the notion of the plurality of worlds in order to explicate whatever topic they were addressing. In so doing, attention was given to the following themes: (1) matter always has been—it cannot be destroyed; (2) matter is organized into worlds; (3) millions of worlds are needed to provide a place of habitation for God's offspring; (4) worlds are created specifically to be inhabited by sentient beings for purposes of progress; (5) worlds are ordered about a great center; (6) worlds are ordered by degree of perfection; (7) worlds are inhabited and organized in kingdoms with their bounds being set; (8) worlds are eternal—they always have been and they always will be; (9) worlds continually form into being and pass out of being; (10) every world has an Adam; (11) the earth is the most wicked; (12) sin exists on every world—therefore all worlds require a redeemer; (13) each system of worlds has its own god; (14) God presides over millions of worlds organized in systems (galaxies?) without end; (15) conditions on worlds differ; (16) resurrected humans will minister to inhabitants of other planets; (17) some worlds are glorified and have become so by the power of the priesthood; (18) the sun is inhabited; (19) stars are not celestialized; (20) there are planets in outer darkness; and (21) there may be communication between worlds.[35]

Virtually all these ideas were developed within a theological context and then further sustained by use of one or more of the teleological, eschatological, analogical, cosmological, or plenitude arguments. For instance, the teleological argument sustained the view that matter is organized into countless worlds, which, in turn, are needed as places of inhabitation of God's offspring, and that the sun therefore must be inhabited; otherwise it would have been created in vain. The principle of plenitude was used to support the idea that not only are God's creations ubiquitous but that every world has an Adam, is covered with sin, and therefore on both points requires a redeemer.[36] Orson Pratt had even earlier invoked the plenitude argument of the Greek atomists and argued that "an infinite number of atoms is requisite to be *everywhere* in infinite space."[37]

Even though the discussion of this pluralistic cosmology appeared in a wide diversity of theological topics by the church's appointed authorities, as a whole the themes are mostly consistent with one another. Only themes (7) and (21) were construed as possibly mutually contradictory: although the bounds of worlds are set, yet there may be communication of some form between worlds. Because Orson Pratt asserted both positions, however, it is likely that his notions of "bound-edness" and "communication" do not refer to the same idea. Also themes (8) and (9) really refer to two different concepts: on the one hand, once worlds have become "eternal" they remain in a celestialized state; while on the other hand, worlds passing "out" of existence does not refer to their annihilation but rather to their "progression" from a degenerate physical state to a celestialized condition. Perhaps the most intriguing theme is (13), which is in keeping with the notion that post-celestialized beings may eventually become gods themselves. Theme (21) has direct bearing on one of several Mormon positions relating to the contemporary scientific program Search for Extraterrestrial Intel-ligence (or SETI) that assumes that communication between techno-logically advanced civilizations is not only possible but inevitable.[38]

Fully two-thirds of the references cited here are attributed roughly equally to Orson Pratt and Brigham Young. Contrary to some who believe that Brigham Young was merely a colonizer and an astute busi-nessman, recent studies, also supported by the evidence presented here, indicate that he was one of the most significant contributors to the theology of Mormonism. Although Orson Pratt's passion since early childhood was astronomy, he, too, along with his brother Parley P. and Brigham Young, was one of the greatest intellects of the early church. Thus it comes as no surprise that these two authorities spec-ulated widely on worlds without number, incorporating the idea into their views on Mormon theology.[39] Though not called to positions of general leadership and authority, other Mormon writers increasingly throughout the nineteenth century used this pluralist cosmology in a wide range of articles, mostly published in church-operated magazines such as the *Deseret News*, the *Millennial Star*, the *Contributor*, the *Ju-venile Instructor*, and the *Young Women's Journal*.

## Mormon Cosmology and the Solar System

Writing in 1888 Mormon geologist James E. Talmage expressed the prevailing scientific sentiment about the universe at large: "The sun, with its family of planets, and their satellites or moons, constitutes the so-called solar system. Wonderful as are these circulating bodies, and great as are their separating distances, this system appears to be but one among many in the boundless fields of space."[40] What precisely then were the various celestial objects? As early as the seventeenth century, astronomers using occultation techniques had shown that the moon does not possess an atmosphere. Despite this scientific evidence the popular press continued to speculate on lunar habitation for several more centuries. With their strong pluralist leanings, Mormons also continued to endorse these fanciful views until roughly mid-nineteenth century, when, along with most Americans, they relinquished the idea.[41]

What of the sun, however? Although Brigham Young and others asserted that stars are not celestialized, the sun itself, some argued, may be inhabited. Mormons and non-Mormons alike often held this view, though by late in the nineteenth century it was conclusively determined that stars are in actuality suns, with conditions that clearly preclude any form of life known to science. Brigham Young himself was an advocate of an inhabited sun as late as 1874, three years before his death. The modern idea of an inhabited sun, at least in its fullest form, dates from the work of William Herschel, discoverer of Uranus (in 1781) and the most famous astronomer of the late eighteenth and early nineteenth centuries. Herschel had argued that the sun, the primary "planet" of the solar system, had an atmosphere with two layers of clouds—the upper layer a luminous shell and the lower layer a planetary cloud needed to protect a solid surface. Herschel developed a theory of matter in which heat, light, various elemental particles, as well as active principles all combined to sustain a cosmic system unifying planets, stars, comets, nebulae, and the sun. In so doing, he achieved a universal link between the development of life and the existence on these many celestial bodies of these material conditions needed to sustain their existence. Then by analogy to the idea of the universality of life prevalent at the time, Herschel suggested that the sun could be viewed as an ultimate extension of inhabited life elsewhere. "The Sun," wrote Herschel in 1795, "is most probably also inhabited, like the rest of the planets, by beings whose organs are adapted to the peculiar circumstances of that vast globe."[42]

While the view of an inhabited sun was developed most fully by Herschel around the turn of the nineteenth century, the Scottish natural philosopher David Brewster, in his popular *More Worlds than One: The Creed of the Philosopher and the Hope of the Christian* (1854) dealing with life throughout the universe, noted that this view of the sun was still acceptable in the 1850s. Brewster's book itself was a rhetorical and polemical response to William Whewell's anonymously published *Of the Plurality of Worlds: An Essay* (1853), in which Whewell shocked his contemporaries by suggesting that pluralists' arguments were utterly misguided, being both scientifically and religiously suspect. Many others jumped into this controversy polarized by Whewell on the one hand and Brewster on the other, including an anonymous critic of

Brewster who charged that Brewster's attitude to scripture was so pedantic that it rivaled the Mormon's propensity for fanciful speculation.[43]

Questions of inhabitation, not only of the sun but also of the planets and other solar systems, was a debate that by mid-century deeply divided many groups—not merely scientists but also theologians, poets, litterateurs, and religious and laypeople. By the early 1870s, however, the more modern view that the sun consists of an incandescent liquid core (subsequently changed to a dense gaseous material), propounded by the German astrophysicist G. R. Kirchhoff, implied that the sun could not support life in any form known to science as suggested earlier by Herschel.[44] Although for a short time after Kirchhoff's work Brigham Young continued to assert the view of an inhabited sun (there almost always exists a time lag in the popular dissemination of virtually any scientific idea), by 1888 the *Millennial Star* carried an article expressing many of these newer ideas, including the concept that stars are identical in chemical and physical composition to our sun.[45] The article went on to suggest that, by analogy to the solar system, star systems in general possess inhabited planetary systems.

Since Greek antiquity, astronomy, as the paragon of all the sciences, had captured the imagination of most people.[46] After all, God was in His heavens and astronomy presented humankind with direct and immediate experience of His presence and edifice. It is no wonder, therefore, that Mormonism, with its theologically relevant interest in plurality of worlds, should also be generally interested in all astronomical matters. Indeed, although he noted that astronomers could not discover specifics of other worlds, Brigham Young believed that astronomy was the greatest of all the sciences.[47] A perusal of Mormon publications, such as the *Journal of Discourses*, the *Millennial Star*, the *Young Women's Journal*, and later the *Improvement Era* and other publications, including church manuals, reveals that essays on astronomy and science appear with consistent regularity. For instance, articles dealing with a variety of astronomical topics, such as the discovery of additional asteroids, the hypothetical inter-Mercurial planet Vulcan, the transits of Venus in 1881 and 1887, the so-called 'Jupiter effect,' space and stars, entropy (the winding-down effect), and, of course, the return of Halley's comet in 1909, were published.

Speculations on the habitability of Mars and Venus had been perennial favorites of pluralists, particularly because, from about mid-century on, the moon and sun had increasingly faded as possibilities for supporting extraterrestrial life.[48] This all changed in 1877, as well as later in 1892 and again 1909, as Mars reached opposition—the point at which the planet is opposite the earth from the sun, closest to the earth, and therefore in the most favorable position for observation. As a result, two major events in our understanding of Mars occurred in 1877. First, the American astronomer Asaph Hall, using the newly acquired twenty-six-inch refracting telescope at the U.S. Naval Observatory, discovered the two tiny moons orbiting Mars. The second event, of far-reaching significance, was the discovery by the Italian astronomer Giovanni Schiaparelli of an extensive system of "canali" on Mars. While the Italian word *canali* is more properly rendered into English as *channels*, suggesting a natural phenomenon, *canali* was almost always rendered as *canals*, implying some sort of human-like workings. The Martian canals described in detail by Schiaparelli were complicated networks of geometrical design. Thus, after continued observations through the 1880s, not only did Schiaparelli's work generate new astronomical interest in Mars but, more importantly, it stimulated in the minds and emotions of people throughout the world excitement about the possibility of extraterrestrial life, not only within our own planetary system but on earth's closest planetary neighbor.

Astronomers across America and Europe continued to add to our understanding of the nature of the Martian surface and atmosphere. Although most scientists remained either publically noncommittal on the issue of the habitability of Mars or skeptical of the possibility altogether, two astronomers in particular—the Frenchman Camille Flammarion, one of the two most prominent celestial mechanists of the nineteenth century, and the American Percival Lowell, scion of one of Boston's wealthiest and most influential families—became strong advocates for the idea that the Martian canals had been constructed by intelligent beings and therefore that life exists on Mars. Schiaparelli himself cautiously supported the notion, but the American Lowell aggressively championed the idea of an inhabited Mars, an idea that captured the popular imagination in staggering ways.

Still, it was the work of the German scientists G. R. Kirchhoff and Robert Bunsen in the 1860s and '70s and the subsequent emergence of the modern science of astrophysics (the practical and theoretical application of laboratory principles of physics to the science of astronomy) that brought new and powerful tools that soon revolutionized the study of the heavens. Already by the late 1870s and early '80s, Mormons were beginning to discuss the new astrophysical developments, wondering what implications these new technological tools would have for the Mormon pluralist doctrine. Writing in 1879 the Mormon George Reynolds observed: "By the spectroscope [astronomers] have demonstrated that the planets . . . and fixed stars . . . are composed largely, if not entirely, of the same elements." Applying an understandable Mormon interpretation, he noted that these celestial objects must "undoubtedly [be] in different states of eternal progression." This being admitted, Reynolds concluded that "if the worlds are composed of the same elements, is it not reasonable to conclude that they are created for the same purpose, to be the homes of humanity, the dwelling places of the sons and daughters of God?"[49]

True enough, Talmage concurred in his *First Book of Nature*, except for the stars, which, although "self-luminous bodies like our sun," are themselves incapable of habitation.[50] Still, argued Talmage, stars "may be the central orb of a vast system of planetary bodies, equal to, and perhaps surpassing our own in grandeur."[51] Thus, whereas the sun and moon had fallen from grace as places of habitation, Talmage's view justified Reynolds's assertion, citing similarities between Earth and Mars already noted, that it has been proven that Mars is the "counterpart of the earth"; that Mars contains oceans, lakes, seas, islands, continents, clouds, snowy poles, and a hot equatorial zone. Invoking the teleological argument and assuming plenitistic reasoning, Reynolds concluded: "Mars is a home of the children of men. To think otherwise would be entirely contrary to the law of probabilities, and still leave unanswered the vital question, for what use was it and the rest of the planets created?"[52]

Along with an active interest among most Americans in the idea of intelligent Martians, the Mormon press reflected interest in and enthusiasm for the idea. Even before Lowell began to contribute to the controversy, the *Millennial Star* editorialized in 1891: "The oceans and continents on our sister planet [Mars] are clearly visible, and the changes that have been observed on the surface of these continents are thought to have been the erection of extensive military works, the building of cities, and, in cases where series of bright flashes have been noticed, an endeavor to communicate with the inhabitants of this mundane sphere."[53]

During the 1890s the church's *Deseret News* kept its mostly Mormon audience abreast of Martian developments, encouraging the attitude "that the question of [Mars] being inhabited is hardly any longer a matter of doubt."[54] Essays by both Lowell and Flammarion were also reprinted in Mormon journals, and issues dealing with direct signal communication with the Martians, the Martian atmosphere, and the habitability of Mars continued to appear until Lowell's death in 1916.

With their new tools, particularly the spectroscope, photometric instruments, and the newer class of big telescopes, by century's end astronomers conclusively demonstrated the chemical and physical uniformity of the various objects within the solar system. This enlivened the pluralist debate, because it showed scientifically that, if most celestial objects differed in degree, they resembled one another in kind. Still, the new instruments also showed that these objects varied enormously in temperature and atmospheric conditions. Some Mormons argued, however, that the planetary atmospheres (e.g., on Mars, Venus, Jupiter) of the solar system may be such that, while receiving only attenuated light from the sun, they could absorb those "vital rays" needed to sustain life. One Mormon author, anticipating a contemporary argument, provided a novel solution to the dilemma implied in these temperature and heat variations on the various planets:

> There is another thing that those scientists who pronounce certain planets unfit for the support of human life do not seem to fully consider. That is, the possibility that men and animals and plants may exist there with constitutions different in some respects from those of this earth, and so able to endure the peculiar conditions existing on those planets. . . . [V]arying situations of planets near to or remote from the sun may be such as are suited to varieties of men, and animals, and plants, similar in some respects but differing in others to the families of their kindred kingdoms on terra firma or in the briny sea of this little globe.[55]

While Americans generally, including Mormons, viewed Mars and other planets of the solar system as inhabited, the Mormon press continued to publish an occasional article devoted to a critical, skeptical view of the inhabitation of any planets other than the earth. In all cases these authors argued that if other planets are inhabited then their beings, who must still be "in the image of God," must possess unusual mechanisms to protect them from the harsh conditions.[56]

## Modern Cosmology and Mormon Theology

Through the nineteenth century Mormons developed the idea of astronomical pluralism to satisfy their understanding of the larger physical universe. In so doing, Mormons nearly perfectly reflected the American cultural milieu and its passion for astronomical pluralism. The idea of worlds without number, however, was also used to express significant and fundamental elements of the newly emerging theology of Mormonism. And thus Mormons developed an interesting and unique cosmology.

To be sure, Mormons explored the implications of pluralism as they speculated on the nature of postmortal existence. But did Joseph Smith and others address a larger purpose by advancing the doctrine of worlds without number? If Joseph Smith provided a framework for astronomical pluralism that extended beyond the concerns of his contemporaries, what purposes did his doctrine serve? Let me suggest the following, admittedly an after-the-fact historical justification for the evolution of his pluralistic cosmology.

Within Mormon theology, as it developed within the Utah church, the concept of the plurality of worlds has implications eventually extending far beyond the idea of multiple inhabited worlds. Fundamentally, the plurality doctrine is wedded to a complex fabric with both theological and religious dimensions. Theologically, astronomical pluralism is a necessary feature of the other forms of Mormon pluralism—wives and gods. Brigham Young later expressed Joseph Smith's pluralism holistically. Speaking on plural marriage before a general conference session on 6 October 1854, Young clarified this radical feature of the restorationist gospel: "The whole subject of the marriage relationship is not within my reach or in any man's reach on this earth. It is without the beginning of days or the end of years; it is a hard matter to reach. We can feel some things with regard to it: it lays the foundation for worlds, for angels, and for Gods; for intelligent beings to be crowned with glory, immortality, and eternal lives."[57] Within Mormonism the sealing of men and women is the underlying condition needed to attain godhood. In turn, God (male/female) propagates spiritual, and eventually physical, progeny that requires, of course, worlds for inhabitation. Thus the complex of pluralism—wives, gods, worlds—established the fundamental basis of nineteenth-century Mormon cosmology.

This view of pluralism also has profound religious significance: within the Mormon concept of the temple pluralism takes on a dimension beyond the theological relationship of these ideas. The temple not only represents the sacred place needed to consummate eternal sealings and blessings but also becomes a microcosm of the universe at large, entailing both symbolic and ceremonial representations of the various heavens of Mormon afterlife. Here pluralism is illustrated "as a program of intense and absorbing activity which [rewards] the faithful by showing them the full scope and meaning of the plan of salvation."[58] Within this scheme the plurality-of-worlds doctrine for Mormonism allows for the completion of the divine creation process. In this sense Mormonism possesses an eschatological orientation that looks toward pluralism in its various dimensions for its ultimate justification.

## NOTES

1. For one theologian's attempt to rationalize this apparent anachronism, see Sidney B. Sperry, *Problems of the Book of Mormon* (Salt Lake City: Bookcraft, 1964), 172–75.

2. The first direct observational data to confirm heliocentrism was the discovery of stellar aberration by James Bradley in 1725 (see chap. 6). Around 1838 and 1839 additional, independent, and conclusive evidence demonstrating the earth's motion in space was uncovered with the discovery of stellar parallax by Friedrich Bessel and others: see N. S. Hetherington, "The First Measurement of Stellar Parallax," *Annals of Science* 28 (1972), 319–25; and O. Struve, "The First Stellar Parallax Determination," in *Men and Moments in the History of Science*, ed. H. M. Evans (Seattle: University of Washington Press, 1959), 177–206, 224–26.

3. Historically, both geocentrism and heliocentrism entail radically different cosmologies. Though not common, pre-Christian views of heliocentrism were available. For a consideration of early Greek cosmology and heliocentrism, see Olaf Pedersen and Mogens Pihl, *Early Physics and Astronomy* (New York: American Elsevier, 1974), 59–67; Stephen Toulmin and June Goodfield, *The Fabric of the Heavens: The Development of Astronomy and Dynamics* (New York: Harper and Row, 1961), 122–27; and William Dibble, "The Book of Abraham and Pythagorean Astronomy," *Dialogue* 8 (Fall 1973), 134–38. Mesoamerican cosmologies are neither geocentric nor heliocentric and therefore represent yet a third possibility: see Michael Coe, "Native Astronomy in Mesoamerica," in *Archaeoastronomy in Pre-Columbian America*, ed. A. Aveni (Austin: University of Texas Press, 1973), 3–31.

4. Orson Pratt, "General Reflections on Eternal Existence," *Millennial Star* 10, no. 12 (1848), 334.

5. There are a number of references in Abraham which deal explicitly with cosmology and pluralism, viz., Facsimile no. 2 (figs. 1 and 2); Abraham 1:31; and Abraham 3:1–17. Although all three references were originally published first in 1842, it appears that Joseph Smith had already understood his Abrahamic astronomy by 1835. In the October 1 entry of the *Documentary History* Joseph Smith is recorded as having written. "This afternoon I labored on the Egyptian alphabet, in company with Brothers Oliver Cowdery and W. W. Phelps, and during the research, the principles of astronomy as understood by Father Abraham and the ancients unfolded to our understanding, the particulars of which will appear hereafter" (see *History of the Church of Jesus Christ of Latter-day Saints*, comp. B. H. Roberts, rev. ed., 6 vols. [Salt Lake City: Church of Jesus Christ of Latter-day Saints, 1948], 2:2860. For the first published references to Abrahamic astronomy, see *Times and Seasons* 3 (March 1842), 703, 705, 719–20; and *Millennial Star* 3 (July/August 1842), 33, 35, 49–53. For an examination of the dates surrounding the production of the Book of Abraham, see Hugh Nibley, "The Meaning of the Kirtland Egyptian Papyrus," *Brigham Young University Studies* 11, no. 4 (Summer 1971), 350–99. Also see Roberts, *History of the Church*, 2:235–36, 286, 348–51. Joseph Smith's discovery and subsequent translation of the Book of Abraham has been discussed extensively: see Hugh Nibley, *Abraham in Egypt* (Salt Lake City: Deseret Book, 1981); and Hugh Nibley's translation of the Abrahamic papyri in *The Message of the Joseph Smith Papyri: An Egyptian Endowment* (Salt Lake City: Deseret Book Co., 1975). For a discussion of Abrahamic astronomy, see H.

Kimball Hansen, "Astronomy and the Scriptures," in *Science and Religion: Toward a More Useful Dialogue*, ed. William M. Hess and Ray T. Matheny (Geneva, Ill.: Paladin House, 1979), 181–96; and R. Grant Athay, "Worlds without Number: The Astronomy of Enoch, Abraham, and Moses," *Brigham Young University Studies* 8, no. 4 (1968), 255–69. For attempts to place Abrahamic astronomy into its indigenous context, see Dibble, "The Book of Abraham and Pythagorean Astronomy," 134–38.

6. Although there exists no significant Mormon hermeneutic tradition that explicates these ideas metaphorically or allegorically, one of the most interesting interpretations of Abrahamic astronomy is given in Hugh Nibley, "Before Adam," in *Old Testament and Related Studies: The Collected Works of Hugh Nibley*, Vol. 1, ed. J. W. Welch, G. P. Gillum, and D. E. Norton (Salt Lake City: Deseret Book, 1986), 67–75.

7. See Hansen, "Astronomy and the Scriptures," 181–96.

8. Joseph Smith, quoted in Frederick Pack, "Natural and Supernatural," *Improvement Era* 11 (January 1908), 182. Also see *History of the Church of Jesus Christ of Latter-day Saints* (Salt Lake City: Deseret Press, 1950), 6:308–9; Sterling M. McMurrin, *The Theological Foundations of the Mormon Religion* (Salt Lake City: University of Utah Press, 1965), 8–11.

9. See Thomas S. Kuhn, "Energy Conservation as an Example of Simultaneous Discovery," in *Critical Problems in the History of Science*, ed. Marshall Clagett (Madison: University of Wisconsin Press, 1969), 321–56; and D. S. L. Cardwell, *From Watt to Clausius: The Rise of Thermodynamics in the Early Industrial Age* (Ithaca: Cornell University Press, 1971).

10. For traditional Mormon accounts of the creation story, see Frank Salisbury, *The Creation* (Salt Lake City: Deseret Book, 1976), 50–109.

11. See McMurrin, *Theological Foundations*, 47–48; and Steve Dick, *The Plurality of Worlds: The Extraterrestrial Life Debate from Democritus to Kant* (New York: Cambridge University Press, 1982), 11–12, 183–84.

12. Anonymous, "Nature," *The Evening and the Morning Star* 1 (October 1832), [40], reprinted from the *Christian Messenger*.

13. Anonymous, "He that will not work, is not a disciple of the Lord," *The Evening and the Morning Star* 1 (November 1832), [47].

14. Anonymous, "Signs in the Heavens," *The Evening and the Morning Star* 2 (December 1833), 116.

15. "The New Year," *The Evening and the Morning Star* 1 (1833), 123; and "New Hymns," *The Evening and the Morning Star* 1 (1833), 141.

16. Oliver Cowdery, "Dick's *Philosophy of a Future State*," *The Messenger and Advocate* 3 (December 1836), 423–25; and "Thomas Dick, *The Philosophy of Religion*," *Messenger and Advocate* 3 (February/March 1837), 461–63, 468–69.

17. Joseph Smith, *The Words of Joseph Smith*, comp. A. F. Ehat and L. W. Cook (Provo: Brigham Young University Press, 1980), 41, 299; and *The History of the Church*, 6:307, resp.

18. Joseph Smith, "A Vision," (1843) in *Revelations of the Prophet Joseph Smith*, ed. L. W. Cook (Salt Lake City: Deseret Books, 1985), 158–66; "God,"

*Times and Seasons* 4 (1843), 367; "Address to the Comet," *Millennial Star* 5 (1844), 48; W. W. Phelps, "The Sky," *Times and Seasons* 6 (1845–46), 895.

19. Parley P. Pratt, *The Millennium, and Other Poems* (1840), 112.

20. Parley P. Pratt, in *Millennial Star* 6 (1845), 10–11.

21. Neptune was not discovered until 1846, while Pluto's discovery had to await the twentieth century.

22. Orson Pratt, "Angels," *New York Messenger* (September 1845), 284, later reprinted as "Mormon Philosophy," *Millennial Star* 7 (1846), 30–31. Occasionally Mormon lay preachers would explore some of these themes: see, for example, "Elder M. Sirrine's Lecture before the Assembly of the Saints," *Millennial Star* 9 (1847), 148.

23. "Diary of Oliver B. Huntington, 1847–1900," Brigham Young University Archives, 168–71; Oliver B. Huntington, "The Inhabitants of the Moon," *Young Women's Journal* 3 (March 1892), 263–64; idem, "Resurrection of My Mother," *Young Women's Journal* 5 (April 1894), 345–47. For a discussion of this entire episode, see Van Hale, "Mormons and Moonmen: A Look at Nineteenth-Century Beliefs about the Moon—Its Flora, Its Fauna, Its Folk," *Sunstone* 7 (September/October 1982), 12–17.

24. Eugene England, ed., "George Laub's Nauvoo Journal," *Brigham Young University Studies* 18, no. 2 (Winter 1978), 176–77.

25. Locke's articles with additional commentary were soon available for wide distribution by W. N. Griggs, *The Celebrated 'Moon Story': Its Origin and Incidents with a Memoir of Its Author* (New York: Bunnell and Price, 1852). Recently, many lunar fantasies have been republished, including Locke's "From 'Great Astronomical Discoveries Lately Made by Sir John Herschel at the Cape of Good Hope (1835),' " in *The Man in the Moone and Other Lunar Fantasies*, ed. F. K. Pizor and T. A. Comp (New York: Praeger, 1971), 190–216. For a thorough discussion of the "moon hoax," as it came to be known, see D. S. Evans, "The Great Moon Hoax, I and II," *Sky and Telescope* (September/October 1981), 196–98, 308–11.

26. The first serious speculations on lunar inhabitation occurred during the seventeenth century by Johannes Kepler, published posthumously as *Somnium* (1634), and by John Wilkins, *Discovery of a World in the Moon* (1638). A recent critical edition of Kepler's book can be found in Edward Rosen, *Kepler's Somnium* (Madison: University of Wisconsin Press, 1967). For the cultural implications of these ideas, see M. H. Nicolson, *A World in the Moon: A Study of the Changing Attitudes toward the Moon in the Seventeenth and Eighteenth Centuries* (Northampton, Mass.: Smith College, 1936); and idem, *Voyages to the Moon* (New York: Macmillan, 1948). Most recently, see John S. Tanner, " 'And Every Star Perhaps a World of Destined Habitation': Milton and Moonmen," *Extrapolation* 30, no. 3 (1989), 267–79.

27. M. J. Crowe, "New Light on the Moon Hoax," *Sky and Telescope* (November 1981), 428–29.

28. Sam W. Taylor, *Nightfall at Nauvoo* (New York: Avon, 1973), 163. In private correspondence, Taylor now believes that Joseph Smith did indeed

believe in moonmen (S. W. Taylor to E. R. Paul, 6 February 1991 [in author's possession]).

29. The fact that speculations about lunar and solar inhabitation saturated Joseph Smith's cultural environment, as well as his views on the plurality of worlds generally, counters the claims of some historians who attribute these views to circumspect magical sources; see D. Michael Quinn, *Early Mormonism and the Magic Worldview* (Salt Lake City: Signature Books, 1987), 130.

30. S. Spalding, *Romance of the Celes* (unpublished MS, Library of Congress, Washington, D.C.). I am grateful to Lester Bush for bringing the existence of this MS to my attention. On the question of the Book of Mormon, see L. Bush, "The Spalding Theory: Then and Now," *Dialogue* 10, no. 4 (Autumn 1977), 40–69; and C. H. Whittier and S. W. Stathis, "The Enigma of Solomon Spalding," *Dialogue* 10, no. 4 (Autumn 1977), 70–73.

31. Parley P. Pratt, *Key to the Science of Theology* (Liverpool: F. D. Richards, 1855), xi. At least eleven subsequent editions of this popular book have been printed. I have used the original 1855 edition.

32. Ibid., 48, 165.

33. The 1979 edition has substantive changes not reflected or justified in either the 1855 or the more widely available 1879 editions.

34. See Orson Pratt, *The Prophetic Almanac for 1845* (New York: Prophet, 1844); Orson Pratt, *The Prophetic Almanac for 1846* (New York: New York Messenger, 1845); J. Dixon, "The Plurality of Worlds," *Millennial Star* 15 (1853), 626–31; and W. W. Phelps, *Almanacs for the Year [1859 to 1864]* (Salt Lake City, 1859 to 1864). For a brief discussion of the role of almanacs in Mormonism, see David Whittaker, "Almanacs in the New England Heritage of Mormonism," *Brigham Young University Studies* 29, no. 4 (Fall 1989), 89–113. Whittaker's thesis that Pratt included astrology because "his intended audience expected it to be there" (99) is ironically sustained by the developments surrounding Pratt's work on Newtonian science from 1850 to 1875 (see chap. 6).

35. See, respectively, *Journal of Discourses* (1) O. Hyde, 2:113, B. Young, 7:2–3, O. Pratt, 14:240, C. W. Penrose, 26:27; (2) B. Young, 13:248; (3) O. Pratt, 1:60, 21:287; (4) O. Pratt, 19:286–87, 21:287; (5) O. Hyde, 1:130; (7) O. Pratt, 3:103, 19:293, 21:233, C. W. Penrose, 26:27–28; (8) B. Young, 7:333, 8:81, 8:200, 10:1, O. Pratt, 19:291; (9) B. Young, 8:81, 14:71, 17:140, O. Pratt, 16:179, 18:293, 19:286, O. F. Whitney, 26:196; (11) B. Young, 10:175; (12) H. C. Kimball, 5:88, B. Young, 14:71–2, O. Pratt, 17:330–31, 18:290, 19:293; (14) A. L., 3:170, B. Young, 1:39–40, 19:50, O. Pratt, 3:103, 14:233–24, E. Snow, 19:328, C. W. Penrose, 21:22, G. Q. Cannon, 26:245; (15) O. Pratt, 24:25; (17) H. C. Kimball, 5:88, O. Pratt 18:297, W. Woodruff, 25:207; (18) O. Hyde, 5:71–2, B. Young, 13:271; (19) H. C. Kimball, 5:88, O. Pratt, 19:291; (20) O. Hyde, 1:130, 5:71; and (21) O. Pratt, 19:291–92, 294, 20:77.

Also see, respectively, *Millennial Star* (2) B. Young, 37 (1875), 221; (3) A. H. Cannon, 56 (1894), 163, O. Pratt, 12 (1850), 68; (4) O. Pratt, 11 (1849), 310; (6) O. F. Whitney, 48 (1886), 483–84; (7) O. F. Whitney, 48 (1886), 482–83; (9) O. F. Whitney, 48 (1886); (12) O. F. Whitney, 48 (1886); (14) O. Pratt,

11 (1849), 234–39; (18) G. Reynolds, 46 (1879), 657–60; and (20) O. Pratt, 12 (1850), 305–6, 309.

Other sources include (3) B. Young, "Addresses," 6 (29 March 1874); (7) O. F. Whitney, *Deseret News* 37 (1888), 509; (10) B. Young, unpublished Discourse (St. George, Utah, 8 October 1854), 7 (12) A. H. Cannon (Brigham Young University Special Collections), 23 June 1889, O. Pratt, *Seer* 134; (13) W. Woodruff, *Deseret News* 5 (14 March 1855), 6; (15) O. F. Whitney, *Women's Exponant* 24 (1895), 10; (16) "Recollections of Orson Pratt," *Young Women's Journal* 10 (January 1899), 22; B. Young, "Addresses" 6 (14 May 1876); (17) O. F. Whitney, *Deseret News* 34 (1885), 374; and (18) C. L. Walker, 1886, B. Young, "Addresses" 6 (14 May 1876).

36. According to Joseph Smith, Christ's redemption is universal, while Brigham Young's "Adam-God theory" suggested other possibilities: see D. J. Buerger, "The Adam-God Doctrine," *Dialogue* 15 (Spring 1982), 14–58.

37. Orson Pratt, "The Kingdom of God," *Millennial Star* 10, no. 20 (1848), 309.

38. See chap. 9 for a complete analysis of this view as it relates to contemporary Mormon cosmology.

39. See G. J. Bergera, "The Orson Pratt–Brigham Young Controversies: Conflict within the Quorums, 1853 to 1868," *Dialogue* 13 (Summer 1980), 7–49; and David Whittaker, "Orson Pratt: Prolific Pamphleteer," *Dialogue* 15, no. 3 (Autumn 1982), 27–41.

40. James E. Talmage, *First Book of Nature* (Salt Lake City: Deseret News, 1888), 261.

41. Quebec, "The Moon," *Contributor* 1 (1880), 195; Newaygo, "A Mighty Telescope," *Juvenile Instructor* 22 (1887), 60; J. Osterman, "The Surface of the Moon," ibid. 30 (1895), 335.

42. William Herschel, "On the Nature and Construction of the Sun," *Philosophical Transactions* 85 (1795), 46–72, 63. Also see Simon Schaffer, " 'The Great Laboratories of the Universe': William Herschel on Matter Theory and Planetary Life," *Journal for the History of Astronomy* 11 (1980), 81–111; and S. Kawaler and J. Veverka, "The Habitable Sun: One of William Herschel's Stranger Ideas," *Journal of the Royal Astronomical Society of Canada* 75 (1981), 46–55.

43. Anonymous, in *Christian Observer* 54 (1855), 40–41. See John H. Brooke, "Natural Theology and the Plurality of Worlds: Observations on the Brewster-Whewell Debate," *Annals of Science* 34 (1977), 231.

44. See A. J. Meadows, "The Origins of Astrophysics," in *Astrophysics and Twentieth-Century Astronomy to 1950*, Part A, ed. Owen Gingerich (New York: Cambridge University Press, 1984), 3–15.

45. P. Grant, "The Stars: Are They Suns?" *Millennial Star* 50 (January 1888): 19–20.

46. While it is frequently suggested that mathematics is the queen of the sciences, since Greek antiquity astronomy had been considered almost exclusively a *mathematical* science until the rise of astrophysics in the nineteenth century.

47. Brigham Young, *Journal of Discourses* 7:2, 2:112, and 4:201.

48. For a thorough discussion of the life-on-Mars issue, see Crowe, *The Extraterrestrial Life Debate, 1750–1900*, 480–546; and William G. Hoyt, *Lowell and Mars* (Tucson: University of Arizona Press, 1976), passim.

49. George Reynolds, "Are the Stars Inhabited Worlds?" *Millennial Star* 41 (1879), 659.

50. James E. Talmage, *First Book of Nature*, 261.

51. Ibid.

52. Reynolds, "Are the Stars Inhabited Worlds?"

53. J. H. A., "Speculations and Progress of Scientists," *Millennial Star* 53, no. 35 (1891), 553.

54. "The Planet Mars," *Deseret News* 45 (1892), 230. Also see "Mars Signal," ibid. 49 (1884), 423; "The Martian Controversy," ibid. 49 (1894), 358; "Ice Fields on the Planet Mars," *Juvenile Instructor* 31 (1896), 562.

55. Anonymous, "Are the Worlds Inhabited?" *Millennial Star* 44 (1882), 437–38.

56. Quebec, "The Planet Venus," *Contributor* 2 (May 1881), 231; R., "Strange Scientific Suggestions," ibid. 1 (November 1879), 40.

57. Brigham Young, *Journal of Discourses* 2 (6 October 1854), 90.

58. Hugh Nibley, "The Idea of the Temple in History," *Millennial Star* 120 (1958), 228–37, 247–49 (247), reprinted as "What Is a Temple?" in Hugh Nibley, *Mormonism and Early Christianity*, ed. T. M. Compton and S. D. Ricks (Salt Lake City: Deseret, 1987), 355–90 (370).

# 6

# Theology in Science:
# The Case of Orson Pratt

The plurality-of-worlds doctrine was not only an essential part of the fabric of Mormon theology as expressed in its quintessential form in the temple but also an excursion into the science of cosmology. Because in the Mormon view cosmology represents far more than its cosmic vision or worldview, it is therefore also a vehicle that allows us uniquely to perceive the relationship of Mormonism as religion to the science of the day. Ultimately, that relationship must come to grips with the perception and use of science among the Mormons during the latter years of the nineteenth century, as they searched for theocratic independence.

Can one talk of a Mormon science, or at least of Mormon theology *in* science? If so, what is the meaning of science for the Mormon experience? Although it seems reasonable that the publication of Charles Darwin's *Origin of Species* in 1859 would dominate religious issues during the latter half of the century, evolution did not pose a serious concern to most Mormons and to the broader dimensions of Mormon intellectual thought until shortly after the turn of the century. Consequently, to assess the theological understanding of science and its perception among Mormons in the later years of the century one must turn elsewhere. More than any Mormon thinker before or since, Orson Pratt (1811–81)—religious leader, scientist, and missionary extraordinaire—exemplified the theologian-scientist who attempted to *combine* his theology with the science of the day. Pratt not only emerged as Mormonism's first scientist, but, perhaps most importantly, he combined his understanding of Mormon theology with an understanding of the physical universe and thus embodied the larger issue of Mormonism and religion.

## Cosmic Theology

Orson Pratt personified the natural theologian with a Mormon flavor: his entire life was dedicated to the vision that true science (natural philosophy) and revealed religion (Mormonism) were in total harmony; therefore, both reflected the presence of the Divine in His works (nature) and in His word (scripture). Conflict between science and religion, as Pratt defined it, was an utter impossibility:

> Whether we engage in this honorable enterprise or not, one thing is certain, the work will be done. Our educational system must be revolutionized—must be re-constructed upon a new and more perfect basis, adapted to a new age—a new era—far in advance of this old. The great temple of science must be erected upon the solid foundations of everlasting truth; its towering spires must mount upward, reaching higher and still higher, until crowned with the glory and presence of Him, who is Eternal.[1]

He was not naive, however, believing that potential conflict between these two great pillars of human experience could and does exist. Only it was due either to apostate religion or incomplete science or both.

Pratt was entirely a self-educated man, from his earliest years possessing a penchant for things scientific and philosophical. Joining the Mormon movement in 1830 at age nineteen, he began studying on his own algebra, astronomy, and other esoteric subjects.[2] Pratt's biographer has suggested that Orson "first encountered [natural philosophy] during the last, declining period of the Scottish Enlightenment while working at Edinburgh in 1840."[3] This seems a reasonable assessment given the fact that, not unlike his Scottish contemporaries, throughout his life Pratt combined an aggressive penchant for science and philosophy into a holistic belief that came to define his cosmic theology.

Many of his ideas on astronomy and related subjects Pratt shared with his friends and like religious believers at numerous public and private meetings. By the Nauvoo period of the 1840s, Pratt was publicly and frequently referred to as "Professor Orson Pratt"; he taught in the "University of the City of Nauvoo," offering courses in mathematics, science, and literature.[4] At the "Nauvoo Lyceum" in November 1842, Pratt lectured on the question "Is there evidence in the works of nature to prove the existence of a Supreme Being?" suggesting that Pratt was fully engaged in natural theology, a form of which he used throughout his life to provide a metaphysical foundation to his understanding of

Mormon theology.[5] Shortly after entering the Salt Lake valley in 1847, Pratt again engaged his passion for teaching, presenting his ideas on astronomy in a series of twelve lectures delivered before a sympathetic audience in Salt Lake City in 1851 and 1852.[6] By 1855 Pratt was regularly lecturing before various schools and "scientific" societies.[7]

Pratt's science was grounded on an ontology that admitted the active presence of God in the universe as material being. That metaphysics, suggested in the scriptural writings and teachings of Joseph Smith and urged by Orson's brother Parley's universal holism, was developed mostly during his early years, particularly in his pamphlets *Absurdities of Immaterialism* (1849) and *Great First Cause* (1851), which together represent Orson Pratt's definitive exposition of his cosmology.[8] Pratt was attempting ultimately to preserve his understanding of Mormon theology. To do so, he needed to postulate an ontology that would justify the existence and omnipresence of the Holy Ghost and to find the physical basis for such a being. The interaction of a physical, material universe with things spiritual was justified on the grounds that the spirit itself was matter—only "more refined." In the words of Joseph Smith, "There is no such thing as immaterial matter; all spirit is matter, but it is more fine or pure."[9] The only physical mechanism allowing for a ubiquitous substance, however tenuous, was the ether of nineteenth-century science. Thus, despite the fact that Pratt talked the language of descriptive astronomy, in the final analysis his approach was guided primarily by philosophy and his interpretation of Mormon theology.

Whereas Pratt argued for a material metaphysics, his philosophical position is justified, ironically, not in the mechanical tradition of modern science associated with the names of Galileo, Descartes, Gassendi, Newton, and Laplace but in the organic worldview rooted in the Hermetic tradition of the Renaissance. This view was also favored in the eighteenth and early nineteenth century by German idealists such as Schelling, Hegel, and Goethe, by the English romantic tradition of Wordsworth and Byron, and by the American transcendentalists Emerson and Thoreau. For Pratt, the animate, active, and psychic presence of a material god was always the ultimate justification for motion, thought, and perception in the universe. In the categories of the Hermetic philosophy, Pratt's monistic metaphysics was thoroughly occult precisely because the materiality of his world possessed self-volition.

In radical contrast to both the occult world of Hermeticism and the rational Aristotelian metaphysics, the mechanical philosophy of nature that came to define seventeenth-century thinking argued for a metaphysics of nature that admitted only two ontological realities—inert matter and motion. The mechanical philosophy had several different interpretations, however. The Cartesian ontology was a plenum (thus admitting no void space) entirely full of matter in various degrees of sizes, while Gassendi's mechanical ontology admitted only inert atoms and void space. As scientist vis-à-vis alchemist, Newton, whose greatest achievement was to synthesize the (Gassendist) mechanical view of reality with a mathematical ontology, added an independent 'force' to the mechanical view of nature.

Descartes, Gassendi, and most other seventeenth-century natural philosophers rejected occult forces as imaginative at best but illicit and dangerous in the extreme. The mechanical ontology argued that things move in the world, not through some mysterious force or interplay between things but rather through the *direct* contact of particles. Thus, by definition, the mechanical world, full of inanimate particles of materiality, caused things to happen by the direct pressure of one particle upon another. Admittedly, in this sense Pratt, like Newton before him, was a (Gassendist) mechanical philosopher:

> There is in reality only one force in the universe, and that is a self-moving force; all the phenomena of the universe are the effects of this self-moving force, either directly or indirectly; and this force always resides in the atoms of matter, and never extends beyond their surfaces; and therefore can only act in the form of pressure, and can never act where the atoms are not present; its effects can only be transferred to a distance through the medium of other matter in the form of pressure, and not in the form of attraction or repulsion which in all cases is absolutely impossible.[10]

In contrast to the mechanical requirement of *inert* particles, however, Pratt's atomistic ontology admitted particles that possessed self-volition. Thus his ontology of nature was a curious mixture of the occult with the mechanical: he accepted the Hermetic emphasis on the animate and active, yet he rejected Newton's idea of force and adopted a mechanical universe entirely full of atoms.[11]

By retaining the active presence of self-moving particles, Pratt both explained the causative nature of reality—things happen because of the direct pressure of one particle on another—and guaranteed the direct primacy of God's presence in the world. Thus, the "all-powerful substance (called God)" holds everything together, specifically the phenomena—such as cohesive attraction, universal gravitation, repulsion, and chemical affinity—that the early Newton had spent so much time trying to explain, using his alchemical ideas. Above all, Pratt argued: "When God performs a miracle by suspending a law of nature, he does so, not by acting at a distance from where the miracle is performed, but by the actual presence of those parts of his essence which are in contact with the materials on which the miracle is performed."[12] In the final analysis, all natural forces, whether mechanical, gravitational, or chemical, are, in Pratt's view, not forces at all but effects of his self-activating universe: "The force which produces these effects is hidden from the view of mortals. A living, intelligent, self-moving force, is the origin of all the motions and laws of nature."[13] In his *Absurdities of Immaterialism* as well as elsewhere throughout most of his life, Pratt argued that, when carried to the extreme in living organisms, these self-active particles become mutually active and emerge into an organic holism.[14]

Other mechanical philosophers, such as Descartes, reduced the Divine to 'first cause' only, or, as in the case with the later Newton, God's primacy was entirely ad hoc and admitted actively only when the world machine wound down sufficiently for the cosmic designer to intervene in an otherwise self-moving, independent system.[15] Thus, while Descartes, Gassendi, Newton, and other mechanical philosophers in the new age that came to define modern science admitted God provisionally or as a last resort, Pratt was fundamentally motivated by a profound belief in an *active* god and therefore constructed an ontology that demanded an active role for the Holy Spirit, all the while retaining his materialistic ontology.

### Scientific Cosmologist

Whereas Pratt's ultimate cosmology has roots in an organic tradition, his science is justified in terms of the prevailing tradition of post-Newtonian science as basically mechanical. Consequently, Pratt engaged the contemporary sciences of astronomy, cosmology, cosmogony, and mathematics in ways that coincided with the received philosophy of science of his day. Equipped to critique the mainstream science by developing new ideas or by correcting what he perceived as misguided notions, in keeping with the prevailing views of the science of his times, Pratt worked as a thorough going, consistent Newtonian.

By the 1870s the many fluids of the eighteenth century had been reduced to two distinct ethers—the electromagnetic and luminiferous ether and the gravitational ether. The purpose of the gravitational ether was to transmit the attractive influence of one celestial body on another. Thus, for example, the moon and earth are held in orbit about one another because of a gravitational attraction exercised between the two. This attraction, however, is really some sort of force that is transmitted across the 225,000 miles of empty space separating these two planetary bodies. But if that space is really empty, how can anything, let alone powerful attractive forces separated by this vast distance, really pull two massive celestial objects together? As in the materialist tradition of post-Newtonian science, much like the ocean is needed in order to transmit currents and the atmosphere is needed to conduct sound waves, so, it was argued, the gravitational ether is needed to transmit attractive forces. In addition to providing the necessary medium needed to transmit waves, the ocean and the atmosphere also can act to retard the speed of the waves. Consequently, by postulating the existence of such an ether, the scientists of Pratt's time also believed that the ether could have a significant retarding influence on all celestial objects—planets, comets, and stars.

Therefore, one of the pressing scientific questions of the nineteenth century dealt with whether and, if so, to what degree, the (gravitational) ether itself affected material objects (e.g., celestial movements). In actuality, in the nineteenth century the problem concerning the viability of the ether was cast in terms of two nongravitational problems: (1) how to explain light and electrical actions by an ether mechanism, and (2) to what extent the ether is dragged along by the matter moving through it. The first problem was avoidable in the sense that physicists simply applied Maxwell's electromagnetic equations and ignored altogether questions about an ether mechanism. Although this is a perfect example of positivist science eschewing metaphysical considerations, ultimately it was Einstein's theory of special relativity (1905) and the subsequent abandonment of Newtonian theory in general that led to the demise of the ether.[16] The second problem, or so-called "ether drag," was unavoidable, because if the ether was in motion, then a different electromagnetic (ether) field would result than if the ether remained at rest.[17]

Whereas mainstream nineteenth-century physical scientists (primarily physicists) concerned themselves almost exclusively with problems entailed in the electromagnetic and luminiferous ether, Pratt represents a minority tradition (mostly astronomers) that concerned itself with the general viability of the *universal* applicability of Newton's gravitation theory. In terms of the gravitational ether, there were two opposing positions. On the one hand, some argued that the planets slid through the ether perfectly, without any noticeable retardation. If this were the case, then, asked others, how could the alleged gravitational attraction between celestial objects pull them together if the ether does not appear to interact with gross matter? On the other hand, if the ether is dragged along with a planet's motion, then wouldn't the planet be retarded in its motion, causing it to slow down and eventually devolve into its central sun?[18]

By 11 November 1854 Pratt had discovered his "long-sought law governing the planetary rotations," which he first announced publically in August 1855.[19] Except for a brief time spent on some specific mathematical interests, Pratt devoted much of the next twenty years of his scientific studies to understanding and promulgating this law. Briefly, the "law of diurnal periods," as it came to be called, relates the period of a planet's rotation with its mass and diameter: "The periods of planetary rotations upon their axes vary as the square of the cube roots of their [mean] densities."[20] At least some Mormons came to hope that "this [planetary law] should indicate to the world, that the Saints do not intend to be behind them in anything. Their rapid progress in the arts and sciences—their system of theology, based on the revelations of heavens—their improved moral and social relations, and thorough system of general education, will soon place them in the van of all other people. The day is not far distant when they will out-strip the world in everything that is great, good, and exalted."[21] Astronomers other than Pratt, particularly the American Daniel Kirkwood, were also searching for a law that would account for planetary rotations.[22]

At the time it was suggested that Pratt's discovery of his "law of diurnal periods" would rank him "with such men as Kepler, Laplace, and others by whom have been discovered . . . some of the general laws which govern the planetary worlds."[23] His "law" was entirely empirical, however—a trial-and-error fiddling with the relevant parameters, numbers, and data of the planets of the solar system. What was missing, Pratt understood thoroughly to his credit, was the underlying mechanism and explanation for this curious relationship: "This remarkable law, connecting periods of rotation with the masses and diameters of the planets, appears to point to some more original law of a higher order of generalization. Such was the case in regard to Kepler's law, connecting the orbital periods of the heavenly bodies with their distances from their respective centers of motion. Newton demonstrated Kepler's law to be a necessary result of the more general law of universal gravitation."[24]

While generously comparing himself to Kepler, one of the greatest scientific geniuses of the human race, Pratt correctly understood that Kepler's three laws of planetary motion, being empirically derived, were fundamentally inadequate as explanation. It was Newton who, in his *Principia* of 1687, provided the mathematical and theoretical justification for Kepler's laws. Equally, Pratt hoped that "providence may raise up a Newton in our day who shall disclose to us the reason why [Pratt's empirical law is valid]." In 1855 he predicted that he himself would be that person and that he would "present a hypothesis which will . . . account for this curious law obtaining in the solar system."[25] Aside from the personality comparisons, Pratt was absolutely correct in his assessment that empirical relationships without a corresponding theoretical understanding are basically incomplete, because a mere description of a relationship by itself does not provide explanation of the phenomenon in question.

During his last mission to Britain in 1877, Pratt published his *Key to the Universe*—easily his most significant scientific book, which provided an explanation of his planetary law and much more—thus himself becoming the latter-day Newton.[26] Pratt's *Key*, however, is far more than an explanation of his earlier law of diurnal periods. The *Key* was intended to provide a comprehensive discussion of the nature of planetary systems and, in particular, to demonstrate the inadequacies, as Pratt conceived them, of the dominant 'nebular hypothesis' of Laplace.[27] Although Newton had applied his idea of universal gravitation to gross matter, such as planetary bodies, neither Newton nor most of his followers applied this notion to the ethereal medium that presumably filled the space between the celestial objects.[28] To correct this state of affairs, Pratt sought "to vindicate the *universality* of the [gravitation] law . . . and to give it that unlimited freedom of action, which the distinguished name, 'universal,' so appropriately and definitely imports."[29] Pratt went on to delineate nine astronomical problems—dealing with such things as the rotation, direction, eccentricity, inclination, and origin of the planets—that, in his judgment, contemporary cosmology was unable to answer adequately.

Pratt's *Key* is premised on two fundamental propositions: (1) a "resisting ethereal medium of variable density" and (2) a "continuous orbital propulsion, arising from the velocity of gravity and its consequent aberrations."[30] The medium or ether (his first premise) has the characteristics of gross matter: it is nonhomogenous, differential, infinite; it is ponderable, therefore it resists motion; and it transmits, and is subject to, the effects of gravity. Because it resists motion (of planetary bodies, in particular), this alone would cause all celestial objects to slow down and eventually come to a complete halt—thus causing the total demise of the universe. In order to compensate for this obviously catastrophic effect, Pratt proposed (his second premise) a physical mechanism, which he called "orbital propulsion" (this is his "key"), that perfectly maintained the orbital velocities and rotations of planets, comets, and even stars.

Orson Pratt (1811–81). Mormonism's first scientist, Pratt became a popular lecturer on astronomical and theological topics. Courtesy of the Special Collections Department, University of Utah Libraries

# KEY TO THE UNIVERSE,

OR A

## NEW THEORY OF ITS MECHANISM.

FOUNDED UPON A

I. CONTINUOUS ORBITAL PROPULSION, ARISING FROM THE VELOCITY OF GRAVITY AND ITS CONSEQUENT ABERRATIONS;

II. RESISTING ETHEREAL MEDIUM OF VARIABLE DENSITY.

WITH

## MATHEMATICAL DEMONSTRATIONS AND TABLES.

BY

## ORSON PRATT, SEN.

SECOND EDITION,
FROM THE FIRST EUROPEAN EDITION.

PUBLISHED BY THE AUTHOR, AND FOR SALE AT THE HISTORIAN'S OFFICE, SALT LAKE CITY, UTAH TERRITORY.

Orson Pratt's *Key to the Universe* (1877). This little book graced the mantles of many Utah homes, providing Pratt's views of the physical laws governing the universe. Courtesy of the Special Collections Department, University of Utah Libraries

His idea is reasonably straightforward and is based by analogy on 'stellar aberration,' discovered by James Bradley in 1725.[31] Bradley had shown that because the speed of light is finite (and constant, about 186,000 miles per second), when the earth is moving tangentially (in the direction T) to its field of observation (F), a telescope through which the observation of a star (S) is being made must be tilted slightly forward (along the line O'S'—i.e., in the direction of the earth's motion) in order to maintain the star's proper position in the field of view (at F). (See figure 6.1.) This compensates for the star's apparent displacement due to the time it takes the star's light to traverse the length of the telescope.[32] Thus, except when the earth is moving directly toward or away from the star, the observer always sees the star in its *apparent* position S'.

Because Pratt was actually interested in the alleged deleterious effects of gravity, and not simply the aberration of star light, he argued that "the effects of aberration in an elliptic [i.e., planetary] orbit . . . will be the same as the phenomena [*sic*] observed in connection with the aberration of light."[33] Consequently, arguing by analogy to Bradley's discovery, Pratt considered that the sun's gravitation is always displaced in the direction of the planet's orbital direction by a distance

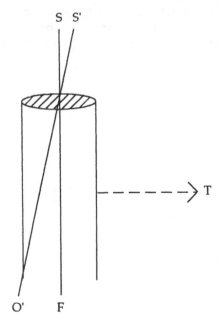

Fig. 6.1—Stellar Aberration

that equally compensates the finite velocity of the sun's gravitational effect on the planet. Therefore, the planet's gravitational attraction to the sun (and vice-versa) is always directed along the tangential line PS', rather than PS. (See figure 6.2.) Since the right triangle PSS' rotates with the planet's motion about the sun, this tangential force (along PS') is due to the gravitational aberration—or, in Pratt's words, 'orbital propulsion'—that accelerates the planet's orbit motion precisely by the amount needed to balance the resistance of the ether to the planet's motion. In this way Pratt believed that planetary systems do not devolve into their central suns but remain stable for long periods of time.[34]

Although Pratt's theory provided ad hoc explanations for the various problems posed in the beginning of his *Key*, it also addressed his law of diurnal periods: why a planet's rotational velocity is merely a function of its density. According to Pratt's view a planet's angular velocity is strictly balanced between two forces—the aberrating force and the resisting force, which, given a planet's density, will increase or decrease the planet's diurnal velocity depending on whether the planet is rotating too slowly or too fast, respectively.[35] Thus—in a slip of logic— Pratt noted that it is not an accident that the earth's daily rotation is slightly more than twenty-four hours.

Ultimately, however, the aim of Pratt's *Key* was not only to demonstrate the inadequacies of Laplace's dominant nebular hypothesis but to provide an alternate mechanism for the evolution of the cosmos. Briefly, under the pressure of his atomistic universe nebulous material

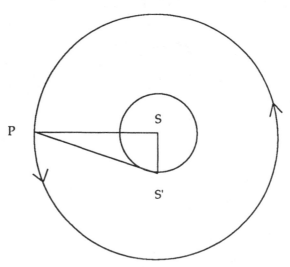

Fig. 6.2—Gravitational Aberration

begins a random accretion process and, under the influence of universal gravitation, begins to rotate, throwing off condensed matter in the form of highly elliptical comets. As these comets enter planetary systems, the ethereal medium about the central sun causes them slowly to adjust their orbits into the planetary plane.[36] In contrast to Laplace's nebular hypothesis, however, Pratt's mechanism does not account for the formations of the most significant astronomical objects—stars. His argument in the *Key* was entirely qualitative, and with the eventual failure of the ether theory, Pratt's idea lost its currency.

Even though Pratt's theory was ingenious, he either ignored, or more likely was simply unaware of, some crucial physical principles—primarily the conservation of energy—because his system allows for work to be done from a region totally devoid of all gross matter, namely the apparent (and unreal) position of the sun (S') along the tangential line of force (PS').[37] Pratt's *Key* also lacked a clear understanding of the optical effects of a differential ether, as well as problems associated with the physics of a medium.[38]

Even though Pratt was well acquainted with the calculus, the most severe methodological problem with the *Key* was that it was couched entirely in simple, algebraic ratios and proportions, not in the differential and integral calculus that had increasingly come to dominate British science after about mid-century. In this regard Pratt was a captive of his earliest training during the 1830s and 1840s, when British science (and with few exceptions, therefore, American science) mostly engaged mathematics using algebraic means and not the higher analysis that had come to dominate Continental mathematics (particularly French and German) from the earliest years of the eighteenth century.

In this regard, over the priority dispute with the French (and indirectly the Germans) concerning whether Newton the Englishman or the German mathematician G. W. Leibniz had first invented the calculus, the British sided with Newton and adopted his fluxion approach to calculus.[39] Mathematically this was a major tactical mistake for the British, because the extremely convoluted nature of fluxion calculus discourages altogether the use of calculus—the most powerful, indispensable tool needed for significant scientific work. Consequently, British mathematics utterly languished, resulting in an elementary algebraic approach to science, while Continental scientists and mathematicians made numerous discoveries and significant progress toward understanding the natural world. Recognizing the deplorable state of much of British science by the 1830s, the natural philosophers John Herschel and David Brewster succeeded in urging the British scientific establishment to relinquish its philosophical—and emotional—hold on Newton and adopt the Leibnizian approach to calculus.[40] By the 1860s British mathematics had begun to reach a level of sophistication that rivaled the development of Continental science. This was particularly evident when the Scottish scientist James Clerk Maxwell first proposed his brilliant mathematical studies that provided a unified theory explicating the fundamentally hitherto disparate phenomena of electricity, magnetism, and light.[41]

Although a contemporary of Maxwell, Pratt generally avoided the use of higher mathematics needed to engage scientific problems at the level proposed by the scientific community—perhaps because he spent considerable time dealing with extremely elementary problems, perhaps because he wrote mostly for a mathematically illiterate audience. Even so Pratt knew a lot of mathematics and was apparently fully capable of sustained mathematical work. He devised a new method for calculating certain algebraic equations of the third and fourth degree (both of which, however, had already received solutions from mathematicians); he also wrote or planned to write texts on the differential and integral calculus, and on the theory of determinants.[42] All of these certainly indicate Pratt possessed considerable mathematical ability. Still, Pratt continued to reflect an outmoded, intuitive, and elementary understanding of science, indicating that the days of amateur contributions to science were waning and would soon be eclipsed entirely.

## Scientific Philosopher

Although Pratt was a Baconian in the sense that he believed in the compatibility of revealed and scientific truth,[43] he rejected Baconianism as a mere collection of facts, as though those facts would somehow, without a theoretical structure, compose themselves into a meaningful mosaic:

> Vague speculations, wild hypotheses, romance, fiction, and every other kindred curse, handed down from the fathers, ought no longer to be considered a part of education. The memories of youthful students ought no longer to be overburdened with isolated facts in a science, when laws comprehending such facts are accessible. Facts may be useful in illustrating laws: but *laws show why the facts exist*. He, therefore, is truly educated in a science, who has understandingly acquired a knowledge of the laws on which the individual facts depend.[44]

In contrast to many, he came to understand fully the importance of *theory* in organizing facts into human knowledge. Perhaps partially for this reason, Pratt was also not a systematizer, a typologist. His philosophical speculations and his scientific efforts, particularly his *Key to the Universe*, provide ample evidence that he was always most comfortable in matters theoretical and abstract. There is a certain irony in the thought of Orson Pratt's methodological approach to matters scientific, however.

With the rise of modern science beginning with Galileo in the seventeenth century and the emergence of the new mechanical ontology of nature most carefully articulated by Descartes, Isaac Newton's greatest contribution was his synthesis of the mathematical and the mechanical traditions into a coherent whole that has been the foundation of modern science ever since. In so doing, Newton, in the emerging tradition of his age of reason with its emphasis on its mathematical language, explicated nature as law. In keeping with the Hermetic philosophers of the Renaissance, however, Newton at the ontological level engaged nature in terms of occult tendencies. As we have seen, this aspect of his science was, by the standards of the seventeenth century, totally out of sync with the mathematical and empirical efforts presented in his *Principia* and *Opticks*.

Orson Pratt also attempted to explicate the laws of nature mathematically. But Pratt, like Descartes, Newton, and many others in the seventeenth century, also continued in that tradition which attempted to explicate the actual, microcosmic workings of nature by constantly speculating on the mechanisms based on its ontology, whereas science, as it continued to develop in the eighteenth and nineteenth centuries, became increasingly positivistic, emphasizing the 'how' of nature and not the 'why.' In this sense, Pratt was most strongly associated with the Romantic and organic vision of reality and therefore correspondingly out of step with the very mechanical tradition that he attempted to emulate. Thus, although Pratt was unaware of Newton's deepest philosophical feelings, Isaac Newton in the seventeenth century and Orson Pratt in the nineteenth became philosophical bedfellows.

Pratt's excursions into mathematics, though modest and actually quite naive, were meant to satisfy his unfulfilled quest for theoretical and mathematical dominance. Pratt understood, in the tradition of Newton and Laplace, that to explicate the mysteries of the universe required thorough knowledge of the language of nature and that that language, at least since the seventeenth century, had become mathematics. Consequently, his *Key to the Universe* was thoroughly saturated with a mathematical and Newtonian veneer.

Advertised for publication in both England and in the east, Pratt's "new theory" followed in the tradition of amateur speculation and naive mathematical manipulation. Despite the popular reception of the *Key* among the Mormon audience, it was hardly more. His Mormon

audience utterly lacked the training to understand the relations between science and mathematics, but Pratt was always frustrated theoretically, because he understood that relationship all too well. To be
sure, Pratt's *Key* reflected a highly imaginative, powerful, and gifted
mind which under different circumstances might have made fundamental contributions to science's understanding of the cosmos. But
untrained in the rarefied atmosphere of obtuse mathematics and physics, Pratt was tragically, but understandably, unable to engage in the
universal conversation among cosmologists.

Richard Proctor, the well-known, highly publicized, and sometimes
controversial English astronomer, understood Pratt's frustration. Although himself frustrated with the British scientific establishment,
Proctor, who engaged the plurality-of-worlds discussion at its highest
levels, ingratiated himself to his Mormon audience by lionizing Pratt
as one of the four greatest mathematicians at that time.[45] (Considering
the array of brilliant mathematicians living at this time, however, it is
utterly baffling what Proctor had in mind.) Some even argued that
Pratt was "one of the seven great astronomers of the world"; again,
Pratt's astronomical contributions were modest at best, and so highly
speculative that they did not merit comment by the scientific community of his day.[46] In a word, among the scientists of Pratt's time, he
was ignored—certainly the ultimate frustration.

### Pratt, Mormonism, and Science

Pratt understood, as few in his religious tradition did, that speculation in itself was insufficient. For Pratt, the way into God's senso-
rium was always dualistic. On the one hand, revelation justified what
in different times was mere speculation. On the other hand, mathematics, which had become the language of nature and justified philosophically since the ancient Greeks and reaffirmed in modern times
in the writings of Newton and his contemporaries and followers, provided the vehicle. Although he espoused a monistic ontology, this
dualist epistemology was for Pratt the ultimate "key to the universe."

In the final analysis, however, Pratt's aim, as in his *Key to the Universe*, was to provide a cosmos that would retain a *physical* basis for
the Holy Ghost and yet still allow for the *physics* needed for the proper
workings of a physical universe with planets, comets, and other celestial
objects orbiting about central stars. Thus, Pratt brought his scientific
penchant and his religious proclivities to bear on the same problem. In
so doing, he found himself trapped in a classic conflict between science
and religion—to wit, if the physics is true, then the religious claim (in

this case the Holy Ghost) should be empirically explicable, while if the religious idea is true, then the science cannot change (which it has, of course). The conflict is only apparent, however, and because Pratt had attempted to unify issues of science and theology in the same cosmic vision, it may be resolved. Pratt's science-religion conflict is not a case of an inherent antagonism; rather, it is the doing of science by incorporating uncritical theological speculation *into* the scientific edifice.[47]

Although Pratt failed to engage science at its highest levels, he understood the need to research, publish, and generally communicate with other scientists using, as he did in his mathematical studies, the journals of his day. Particularly in mathematics, however elementary, Pratt succeeded.

Pratt was not the only influential nineteenth-century Mormon who was centrally interested in the scientific enterprise. Whether for theoretical purposes, as with Pratt, or more mundane reasons, Mormons used science in constructive ways, particularly as they encountered the challenges of an arid climate. Mormons engaged science not only in such esoteric and theoretical matters as astronomy, plurality of worlds, and cosmology generally, but also in matters dealing with agriculture, irrigation techniques, and other practical sciences and technologies.

Despite the fact that Mormons found themselves in a United States territory in Utah, and thus subject to federal law, Mormonism was also highly theocratic, dominated in principle, if not in fact, by prophets and apostles. Consequently, as a result of the use of science among Mormons, for both theoretical (theological!) and practical purposes, Mormonism had fully embraced science—at least in principle. To be sure, there were areas of science (particularly after Darwin) and Mormons who remained mutually suspect of the scientific enterprise. But in a general sense, Mormons, like most Americans generally, mostly confused science with technology and received the science of the day openly. Nowhere is this more evident than within the highest ruling quorums of church governance, where, following Pratt's death in 1881, there emerged a number of powerful general authorities who were themselves trained in science or who adopted a scientism in their theological writings.

Pratt himself perhaps best exemplifies the Mormon penchant for confusing science with some form of objective truth. To be sure, in the final analysis Pratt refrained from fully embracing an epistemological realism. But his repeated efforts at using science to construct a cosmology in harmony with his Mormonism underscores the view that Mormons too frequently and too uncritically accepted the findings of science. Reinforced by Mormonism's emphasis on a philosophical realism, this view encouraged a dangerous liaison between science and religion urging an unhealthy sympathy with natural theology.

### NOTES

1. Orson Pratt, *Deseret News* 22 (1873), 586.

2. Elden J. Watson, comp., *The Orson Pratt Journals* (Salt Lake City: Elden J. Watson, 1975), 91, 164; "University of the City of Nauvoo," *Times and Seasons* 3 (1841/42), 631.

3. Breck England, *The Life and Thought of Orson Pratt* (Salt Lake City: University of Utah Press, 1985), 289–92.

4. "University of the City of Nauvoo," *Times and Seasons* 2, no. 20 (16 August 1841), 517; Watson, *The Orson Pratt Journals*, September 1842, 184; "William Clayton Journal, 7 Jun 1847," in Kate B. Carter, *Heart Throbs of the West*, 7 vols. (Salt Lake City: Daughters of the Utah Pioneers, 1939) 6:267.

5. *Wasp* 1, no. 30 (12 November 1842), [3–4].

6. Orson Pratt, "Astronomical Lectures," *Deseret News* 2, nos. 4, 5, 7, 8, 10–12, 15–18 (1851/1852), reprinted in Orson Pratt, *Wonders of the Universe*, comp. N. B. Lundwall (Salt Lake City: N. B. Lundwall, 1937), 1–156. Lecture number 12 was given, but not printed in either the *Deseret News* or *Wonders of the Universe*.

7. "Notice," *Millennial Star* 17 (1855), 233, 732.

8. See Joseph Smith's 1833 and 1843 revelations in D&C 93:29 and D&C 131:7–8, resp.; Parley Pratt, *The Millennium and Other Poems, to Which Is Annexed 'A Treatise on the Regeneration and Eternal Duration of Matter'* (New York: W. Molineux, 1840), 105–48; Orson Pratt, *Absurdities of Immaterialism; or, A Reply to T. W. P. Taylder's Pamphlet, Entitled "The Materialism of the Mormons or Latter-day Saints Examined and Exposed"* (31 July 1849), 32 pp.; and idem, *Great First Cause; or, The Self-Moving Forces of the Universe* (1 January 1851), 16 pp. Pratt's commitment to a Newtonian universe was already evident in his 1845–46 almanacs.

9. D&C 131:7.

10. Orson Pratt, "The Holy Spirit," *Millennial Star* 12 (1850), 328.

11. See Orson Pratt, "Mormon Philosophy," *Millennial Star* 6 (1845), 157–59, 173–75; 7 (1846), 29–31, esp. 174.

12. Pratt, "The Holy Spirit," 326–27.

13. Orson Pratt, "The Pre-Existence of Man," *The Seer* 1, no. 7 (July 1853), 97–104, esp. 101–2.

14. Pratt, *Absurdities of Immaterialism*, 20–22; and idem, "The Pre-Existence of Man," 103. This view is suggestive of the twentieth-century theory of 'emergentism,' which suggests that inert particles, when organized into chemically active entities, emerge into *biologically* living organisms.

15. For a discussion of the active role of God in Newton's universe and his disagreement with most other mechanical philosophers over this issue, see David Kubrin, "Newton and the Cyclical Cosmos: Providence and the Me-

chanical Philosophy," *Journal of the History of Ideas* 28 (1967), 325–46; Gale Christianson, *In the Presence of the Creator* (New York: Macmillian, 1984), 549–51; and A. Rupert Hall, *Philosophers at War: The Quarrel between Newton and Leibniz Correspondence* (New York: Cambridge University Press, 1980), passim.

16. Einstein's idea of special relativity (1905) was not motivated by the null effects of the 1881 and 1887 Michelson-Morley experiments; see Gerald Holton, "Einstein, Michelson, and the 'Crucial' Experiment," *Isis* 60, no. 2 (Summer 1969), 133–97.

17. See William Berkson, *Fields of Force: The Development of a World View from Faraday to Einstein* (New York: John Wiley, 1974).

18. Hugo von Seeliger, "Ueber das Newton'sche Gravitationsgesetz," *Astronomische Nachrichten* 137 (1894), 132–33; and K. Laves, "On Some Modern Attempts to Replace Newton's Law of Attraction by Other Laws," *Popular Astronomy* 5 (1898), 513–18. Among the most prominent astronomers who concerned themselves with these questions were the Germans Carl Neumann and Hugo von Seeliger. Along with the Dutch astronomer J. C. Kapteyn, Seeliger developed the science of statistical cosmology, the most influential approach to our understanding of the sidereal system developed before the 1920s, without regard to the metaphysics using the tools of fairly sophisticated mathematics. See Erich Robert Paul, *The Milky Way Galaxy and Statistical Cosmology, 1890–1924* (forthcoming from Cambridge University Press).

19. "Astronomical Discovery of the Law of Planetary Motion," *Deseret News*, 1 August 1855, 165; the complete article was published in ibid., 8 August 1855, 172. Also see Scott G. Kenney, ed., *Wilford Woodruff's Journal, 1833–1898*, 6 vols. (Salt Lake City: Signature Books, 1983–86), 11 Sep 1855, 4:334. Also see Orson Pratt, "Law of Planetary Rotation," *London Times*, 21 August 1857.

20. Orson Pratt, *Key to the Universe; or, A New Theory of Its Mechanism* (1877; 1879), chap. 12, part 118. The manuscript of the first 105 sections of his *Key*, called "A New System of the Universe, Founded upon a Constant Propulsion and Resistance," is located in the Orson Pratt Family Collection, Church History Department, LDS Archives.

21. "Professor Pratt's Discovery," *Millennial Star* 17 (1855), 792.

22. See Ronald L. Numbers, "D. Kirkwood's 'Law of Planetary Rotations,'" in Ronald L. Numbers, *Creation by Natural Law: Laplace's Nebular Hypothesis in American Thought* (Seattle: University of Washington Press, 1977), chap. 4; and Stephen G. Brush, "Looking Up: The Rise of Astronomy in America," *American Studies* 20 (1979), 50.

23. "Professor Pratt's Discovery," 792.

24. Orson Pratt, "Law of Planetary Rotation," *Millennial Star* 17 (1855), 797.

25. Pratt, "Law of Planetary Rotation," *Deseret News*, 8 August 1855, 172.

26. Pratt, *Key to the Universe*. For additional insight into Pratt's *Key*, see D. Skabelund, "Cosmology on the American Frontier: Orson Pratt's Key to the Universe," *Centaurus* 11, no. 3 (1965), 190–204; England, *The Life and Thought of Orson Pratt*, 272–78; and David Whittaker, "Orson Pratt: Prolific Pamphlet-

eer," *Dialogue* 15, no. 3 (Autumn 1982), 37–45. Also see John C. Greene, "Some Aspects of American Astronomy, 1750–1815," *Isis* 45 (1954), 339–58.

27. See Numbers, *Creation by Natural Law*, passim.

28. Kubrin, "Newton and the Cyclical Cosmos," 325–46.

29. Pratt, *Key*, preface.

30. Ibid., title page.

31. Anton Pannekoek, *A History of Astronomy* (New York: Barnes and Noble, 1969), 289–90.

32. Today, differentiation is made between "diurnal aberration," caused by the earth's daily (or diurnal) rotation, "annual aberration," caused by the earth's yearly orbit about the sun, and "secular aberration," caused by the motion of the solar system through space.

33. Pratt, *Key*, chap. 3, part 40.

34. Ibid., chap. 3, particularly parts 35–42.

35. Ibid., chap. 12, part 119.

36. For a discussion of additional problems, see Skabelund, "Cosmology on the American Frontier," 197–98.

37. Pratt's assumption that stars act gravitationally on their planetary bodies as though from an imaginary point in space, which has no *physical* basis whatsoever, was not altogether unlike Ptolemy's second century A.D. geocentric cosmology that also required an imaginary point in space, called the 'equant,' about which all celestial objects rotated at uniform speeds. In 1543 Ptolemy's ad hoc equant was rejected by Copernicus, in the latter's heliocentric cosmology, as scientifically and philosophically untenable.

38. Pratt, *Key*, chap. 13, parts 125–30.

39. See Hall, *Philosophers at War*, passim, and Westfall, *Never at Rest*, 698–780.

40. See A. D. Orange, "The British Association for the Advancement of Science: The Provincial Background," *Science Studies* 1 (1971), 316–29; and Jack Morrell and Arnold Thackray, *Gentleman of Science: Early Years of the British Association for the Advancement of Science* (New York: Oxford University Press, 1981).

41. See Berkson, *Fields of Force*, passim, and Christa Jungnickel and Russell McCormmach, *Intellectual Mastery of Nature: Theoretical Physics from Ohm to Einstein: The Now Mighty Theoretical Physics, 1870–1925*, Vol. 2 (Chicago: Chicago University Press, 1986), passim.

42. See Edward R. Hogan, "Orson Pratt as a Mathematician," *Utah Historical Quarterly* 41 (Winter 1973), 59–68.

43. See George Marsden, *Fundamentalism and American Culture: The Shaping of Twentieth-Century Evangelicalism, 1870–1925* (New York: Oxford University Press, 1980), 55–62.

44. Pratt, *Millennial Star* 35 (18 November 1873), 722; reprinted in *Deseret News* 22 (1873), 586 (italics added).

45. Orson F. Whitney, *History of Utah*, 4 vols. (Salt Lake City: George Q. Cannon and Sons, 1904), 4:29.

46. "Astronomical Lectures," *Deseret News* 29 (18 January 1871), 588.

47. For an analysis of comparable problems in nineteenth-century geology and American science, respectively, see Charles C. Gillispie, *Genesis and Geology: The Impact of Scientific Discussion upon Religious Beliefs in the Decades before Darwin* (New York: Harper and Row, 1959); and Herbert Hovenkamp, *Science and Religion in America, 1800–1860* (Philadelphia: University of Pennsylvania Press, 1978).

# 7

## Science in the Church Hierarchy

That life exists throughout the solar system only reaffirmed in the minds of Mormons the belief that world systems were scattered throughout the universe. Through the nineteenth century the notion of worlds without number had evolved into a relatively coherent idea that had become the cosmological basis of Mormon theology. To give it an eschatological dimension the whole idea had been codified theologically in the temple ceremony. Except possibly for Orson Pratt, who was Mormonism's first notable scientist and an amateur contributor to the field of astronomy, those who speculated on this notion were mostly religionists. Moreover, the close association of numerous scientific ideas with many Christian sects during the later years of the nineteenth century, as well as the idea that technology was contributing to the enormous progress that Americans experienced during this period, affirmed in the minds of Americans generally the extraordinarily positive image of science.

### The Emergence of a Mormon Scientism

Although this American love affair with science as technology was to wane somewhat by the middle years of the twentieth century, it was deeply reflected by a significant attitude within the Mormon hierarchy. In the last year of Brigham Young's tenure, self-trained historian and theologian B. H. Roberts had been called into the Seventy Quorum, where, although not a professional scientist, he made significant contributions to Mormon theology generally and to Mormon cosmology specifically. During the first several decades of the twentieth century four professionally trained Mormon scientists—James E. Talmage, John A. Widtsoe, Joseph F. Merrill, and Richard R. Lyman—were called as apostles. Talmage (geology), Widtsoe (chemistry), Merrill (physics), Lyman (engineering), and Roberts treated science generally and, in the case of Widtsoe and Roberts, cosmology and astronomical pluralism specifically in many of their writings.[1]

The fact that the church called four scientists into the Quorum of the Twelve suggests that the church may have been expressing the progressive view of science generally held among Americans. Whether the church was indeed reflecting this positive attitude or whether these scientist-scholars were simply strong advocates for a positive scientism is, in the final analysis, however, not relevant. What is historically sure is that all of these men contributed to sustaining a positive image of science among Mormons primarily by virtue of their professional stature as scientists as well as through their own understanding of science and religion as reflected in their many speeches, sermons, and public writings. The result was the development of three basic, though sometimes conflicting, trends.

First, with the elevation of these scientist-scholars to church prominence, a broadly scientific attitude emerged within the highest levels of church governance. All five of these men used ideas from the larger cultural-intellectual climate of their day to enrich and expand the church's understanding of its theology and its relationship to both science and scholarship generally.[2]

Second, through most of the nineteenth century church members, as well as most Americans generally, saw the achievements of technology as the embodiment of science. In very broad terms, science may be pursued as "pure" or as "applied." As "pure," science is primarily philosophy, a quest for understanding; as "applied," science is technology, or the embodiment of action. Before science can be applied, however, it must be understood as explanation. This confluence of ideas tended to confuse the inventions of technology with their underlying scientific principles. Technology, particularly prior to the First War, was in fact largely benign and, therefore, did not threaten most people's values and beliefs. Moreover, in the minds of most Americans science and technology had become equated with progress. Consequently, the popular image of science tended to remain quite positive.[3] Mormonism's view of science almost perfectly reflected these nineteenth-century attitudes.

Third, as we have noted, the image of science was deeply enhanced by the church's theological underpinnings. Only gradually, after Darwin presented his ideas on evolution in 1859, did some groups in

America begin to distrust science. This nascent antiscientific view became particularly prominent among biblical literalists and Christian fundamentalists. Mormons themselves did not remain insulated by the evolutionary controversy that slowly raged after Darwin into the early decades of the twentieth century. By the beginning of the twentieth century, dispute among some of the church leaders, particularly between Joseph Fielding Smith and B. H. Roberts over the issue of the existence of pre-Adamic life as well as a "modernist" crisis focusing around evolutionary ideas at the church's Brigham Young University, led to some acrimonious exchanges.[4] Although a certain anti-intellectualism and distrust of science eventually emerged among some Mormons in the early decades of the present century, Mormons generally continued to court science and express a positive view toward science well into the twentieth century.[5]

Among their many church books and manuals Roberts's *Joseph Smith: The Prophet-Teacher* (1908), Talmage's *Articles of Faith* (1899) and his *Jesus the Christ* (1915), and Widtsoe's pre-apostolic *A Rational Theology* (1915), which was deeply embedded in a positive scientism, all reflected the church's confidence and ease with scholarly matters. Through their writings and lectures these influential men contributed many contemporary scientific ideas to the writings and beliefs of the church body. For instance, both Talmage and Widtsoe consistently argued for a geological (old-earth) vis-à-vis a biblical literalist (6,000 year) creation, and both were less suspicious of a quasi-Darwinian, or at least a progressive, interpretation of life than most others. Widtsoe, in his important and widely read *Joseph Smith as Scientist* (1908), argued that the teachings of Joseph Smith "were in full harmony with the most advanced scientific thought of today, and that he anticipated the world of science in the statement of fundamental facts and theories of physics, chemistry, astronomy and biology."[6]

## B. H. Roberts on Mormonism and Cosmology

Of the four apostles, Talmage, Widtsoe, and to a lesser degree Merrill used their scientific training forcefully. Yet it was the slightly older nonscientist B. H. Roberts (1857–1933) who was most inclined to weave Mormonism and science into a systematic and coherent system.[7] After the death of Orson Pratt, Roberts became the church's leading defender and apologist and, according to some, the premier Mormon intellectual of his day.[8] His literary output was prodigious. He eventually authored fifteen books, including the six-volume *Comprehensive History* of the church and two biographies, as well as twenty-three manuals and study courses. He also published over 300 articles and reviews in a wide variety of periodicals. His understanding of social, scientific, and literary issues was broad and sensitive. In his official church manual, *The Seventy's Course in Theology*, published in five consecutive years from 1908 through 1912, Roberts treated a range of scientific issues, including the concept of natural law, the new astronomy, and evolutionary studies. In most of his writings Roberts frequently cited the works of relevant non-Mormon sources, including standard science texts as well as the polemical works of Andrew White's *History of the Warfare of Science with Theology* and Henry Draper's *Conflict between Religion and Science*, often responding to and elaborating upon issues dealt with in the literature.

Roberts's most substantial work, however, was not his church books but his summa theologia, entitled "The Truth, the Way, the Life: An Elementary Treatise on Theology," written during the last decade of his life and finished for publication in 1928. In this, Roberts blended a lifetime of views on religion, philosophy, Mormonism, and science into a sort of systematic theology. This treatise was intended to climax his doctrinal views as the *Comprehensive History* completed his historical work.[9] The central focus of his work dealt with Christ in terms first of "the truth," second of "the way," and finally of "the life." It was a treatise on the nature of humankind and the cosmos standing in relation to God as both creator and savior. In it Roberts dealt extensively with science and its relationship to religion. Although the treatise itself is not strikingly innovative in particulars, the depth and forcefulness with which Roberts argued for his version of a Mormon theology coupled with modern secular learning marked it as a unique contribution to Mormon religious and intellectual thinking. As it turned out, parts of his treatise were unacceptable to several members of the Quorum of the Twelve. Even after much discussion and some subsequent revision, the troublesome issues were never resolved to the satisfaction of the disputants.[10] Roberts possessed a rare intellect, and Mormonism probably lost when the manuscript failed to meet their approval and consequently was never published.

In some respects Roberts's views represent the culmination of ideas on natural law and science as expressed in the writings of Talmage, Widtsoe, Merrill, and Mormon scholars Frederick J. Pack, Nels L. Nelson, and others.[11] On the scientific side, he treated extensively issues dealing with astronomy (including the plurality of worlds) and evolution, because these sciences deal most directly with the origin and nature of humans and the universe. Although he treated the secular literature broadly, Roberts also recognized the need to read his sources discriminately, "not accepting either all the premises laid down, or the conclusions reached." In his third chapter Roberts briefly commented on the nature of time, space, matter, and force—what he calls the "building stones of knowledge": "I shall not attempt any discussion of the 'reality' of them at all: shall only deal with such definitions and treatment of them as will make clear what may be presented as the general sum of man's partial knowledge of the solar and sidereal systems that make up the universe in which he lives."[12]

Although Roberts's discussion of these ideas is mostly definitional and not critical, he clearly recognized some significant issues and, in fact, was relatively current in his thinking. For example, noting the nineteenth-century views on the indestructibility and nonconvertability of both matter and energy (force) discussed earlier, Roberts, under the subheading "Twentieth Century Advancement in Physics," noted:

> It has occurred to me that some of our more recent writers and students may take exception to the matter as here set forth. . . . Some of our present day professors hold that the principle of the 'indestructibility of matter' has proven to be 'definitely invalid'; and it is now sometimes held that a definite portion of matter has entirely disappeared as a distinct, separate entity of any system, *energy taking its place*. That is, matter changes into radiant energy, and vice-versa.[13]

Citing Robert S. Millikan, the dean of early twentieth-century American physics, and Albert Einstein as his sources, Roberts concluded that "matter has not been dissolved into 'nothing'. . . . There has been no break in the continuity; something has existed all the while, and the old truth on the conservation of matter and force has not in reality been changed, but emphasized." Noting that "all spirit is matter" (D&C 131:7) and that "there are many worlds that have passed away" (Moses 1:35, 38), Roberts emphasized that the "making and unmaking" of worlds conforms well to this doctrine of the changeability (but not destructibility) of matter-energy.[14]

In subsequent chapters Roberts discussed contemporary views of the nature of the solar system and the sidereal system (Milky Way in contemporary parlance). Roberts was certainly current in his discussion. For example, during the 1920s a major conceptual revolution in astronomy and cosmology had occurred, subsequently dubbed the "second astronomical revolution," in which American astronomers Harlow Shapley and Edwin Hubble were among the principal theorists. They showed that (1) the sun and its solar system are not centrally located in the sidereal system (Shapley); (2) the Milky Way system is about ten times larger than the stellar system previously propounded

(Shapley); (3) there exist external clusterings or galaxies of stars (Hubble); and (4) the universe is expanding (Hubble). Although this new cosmology of the universe has been subsequently modified in important ways, it has received increasing assent from the astronomical community since 1930.[15]

On this point, Roberts was representing the cutting edge of modern astronomical research. But the concern that was foremost in Roberts's mind dealt with the *meaning* of the cosmos. Now that astronomers had demonstrated the extent of the physical universe, what, asked Roberts rhetorically, was the meaning of it all? In a major address before the American Association for the Advancement of Science in 1926, F. R. Moulton, professor of astronomy at the University of Chicago, noted the most recent discoveries of the larger dimensions of the Milky Way (Shapley) and the existence of exterior galaxies (Hubble) and suggested that "it is not improbable—it is in fact probable—that a majority of [stars] have planets circulating about them, as our earth revolves about the sun. It may be that a fraction of [stars], perhaps in a hundreds of millions, are in a condition comparable to that of the earth, and that they support life!"[16] Roberts, always sensitive to such ideas, not only digested Moulton's point, but went on to note Moulton's argument in support of a fundamental premise of science that suggested a deeper meaning to the universe:

> The impressive thing to the astronomers is not the magnitude of the galaxy, nor the long periods of time during which stars exist, nor the tremendous forces of nature; but the most impressive thing is that all this vast universe which we have been able to explore is found to be orderly. The orderliness of the universe is the foundation on which science is built. It is the thing that enables us to understand the present, to look back over the past, and to penetrate the remote future.[17]

Citing additional sources, including Draper, White, and the American intellectual John Fiske, Roberts emphasized that "order" or the "reign of law" is a sine qua non for understanding the nature of the universe. This was not simply a basic premise of science, but, in Roberts's view, a rare insight into the actual operations of the universe. Despite Roberts's earlier caveat about our inability to know finality, on at least this issue, Roberts argued, science had penetrated to the core of physical reality. Roberts understood that even though scientists were compelled to hold this assertion as a fundamental premise for the doing of science, it is still a statement of belief and therefore metaphysical. But Roberts's religious commitments furnished the added assurance that *this* issue indeed is ontologically grounded.

If law and order are supreme in the physical universe, what, asked Roberts again, is the causative agent of the material universe? Assuming the nondestructibility of matter-energy, Roberts argued that "the universe itself is uncaused since it always existed, and is all that is, including all forces whatsoever, as also all intelligences, or mind—it is the sum of existence.' "[18] Similarly, it was not because scientists had somehow demonstrated this assertion but rather, argued Roberts, because Joseph Smith had revealed that (1) because all spirit is matter and spirit is indestructible, matter too must be without beginning, and (2) there never was a time when something material (spirit, etc.) did not exist. Therefore, Roberts also rejected the Aristotelian notion of a "first cause," suggesting contrariwise the idea of an "eternal cause." Furthermore, "this 'first cause' idea involves us in the whole argument of the designer of the universe, a designer that at once is outside of and transcends the universe."[19] Rejecting this traditional Christian view of God and carrying this heresy into his theology, Roberts claimed God is both "in time" and "historical" and, therefore, must be existentially conditioned. Ultimately the purpose of his argument is to convince us that science and Mormonism are reasonably compatible—that, exercising proper caution, we can rely on the statements of science to cast additional, but secondary, light on the statements of "revealed" religion.

Having completed his discussion of some basic scientific and methodological concepts, as well as having introduced various views of science on the nature and origin of the physical universe, Roberts commenced an examination of the plurality of worlds.

> We have ascertained from . . . various authorities upon astronomy that it is possible and even probable that the suns which make up our galaxy—our universe—have circulating about them groups of opaque worlds, even as our sun has nine such worlds moving about him in their respective orbits. But is it true that each of these suns of the sidereal system, or even a considerable number of them, has a like group of planets to which it is the center of gravitation, and from which these planets receive light and warmth and vital force, resulting in life such as we know it on our own earth?

Though Roberts was definitely committed to a plurality of worlds, he realized the speculative nature of the *scientific* question whether in fact there exist worlds without number: "The answer must necessarily be that this is not definitely known, and hence scientists in astronomy speak with caution, and only say that it may possibly be so. It may even be probable. But science can speak with no positive assurance on this subject, because really scientists do not know."[20] Using already familiar philosophical arguments, Roberts asserted that (1) by analogy to the sun's system, stars must be centers of solar systems whose planets must contain sentient, intelligent life; (2) it would be a waste of stellar energy if the sun supported the only inhabited planet in the cosmos; and (3) if there were no other inhabited planets, then what possible purpose would the universe serve? For instance, continued Roberts, the earth would be utterly without purpose if not created and made for humans; everything in and on the earth would ultimately be wasted and have no meaning, if not for humankind. These arguments—by analogy, of plenitude and of teleology—are, in Roberts's view, compelling.

Roberts did not stop with asserting the simple view of a plurality of worlds. Tacitly accepting the nineteenth-century belief in the inevitable progress of humankind, he felt inclined to suggest certain personal and civilizational characteristics that extraterrestrials must possess:

> And there may have been developed also higher and mightier intelligences than any that have been developed on our earth. If such intelligences do exist, in other worlds, may we not enter upon the same line of reasoning from what we know, apply the principle we have been following to the social and sympathetic and moral qualities as connected with these higher intelligences? This we know in respect of the inhabitants of our own world, that higher intellectual life and higher states of civilization produce exalted moral feelings, resulting in higher states of righteousness and love of truth and sympathy for fellow men, leading to desire for the uplift of those less highly developed, and thus is produced among our own earth-people a desire to restrain the strong and vicious by laws . . . and to uplift and better the conditions of the lowly and undeveloped.[21]

Again, by analogy to human conditions, Roberts argued for altruistic and progressive conditions for extraterrestrials. Although Roberts served as a chaplain during the Great War, apparently his experiences and reflections were still not sufficient to embitter him on the question of the human condition. It would perhaps take another world war to convince him that intelligent life, at least of the variety we are all familiar with, is capable of incalculable suffering and tragedy.

Roberts also provided scriptural support for his view of universal, ennobled intelligent life. Noting that Enoch spoke of worlds without number, Roberts quoted the scriptures to reveal a pluralism of life: "I [God] also created [worlds] for mine own purpose. . . . Adam, which is many" (Moses 1:33–34). "From the last statement," wrote Roberts, "it appears that Adam is a generic name, that there are many Adams carrying the significance perhaps of being first placed on the creations of God."[22] But the heavens, Enoch mentioned, are many, and they are numbered only unto God (Moses 1:35). These worlds are peopled by beings which, in the parlance of the book of Abraham and the Doctrine and Covenants, are described as "intelligences," and which, in Roberts's view, are necessary vis-à-vis contingent beings. In the scope of astronomical understanding, these intelligences refer to "other and older world systems." "Be it remembered here," Roberts wrote, "that these kingdoms and the inhabitants thereof are the kingdoms of the space depths in the universe, all the worlds, and the world systems, and by the word of God they have their inhabitants."[23]

Roberts was an extremely positive advocate for a full-fledged scientism, the view that science could provide in large measure an understanding of both the moral and physical universe. His use of science to substantiate not only his understanding of Mormonism but also his vision of a moral life was reasonably certain. He tried to remain current in matters scientific and, in large measure, succeeded. He was willing to defend science as a legitimate means epistemologically for acquiring knowledge. Although his enthusiasm for science at times could be explosive, he also recognized, with the sensitivity of the philosopher, the importance of being constructively critical of the scientific enterprise. Altogether his caution was tempered with a genuine respect for science. Finally, Roberts succeeded in weaving the two forms of human understanding—religion and science—into a richly colored fabric.

But Roberts also recognized something more. A less critical mind would likely conclude that science, epistemologically, could easily be used as the foundation for the various claims of religion, and thus religion would logically be made irrelevant. In his chapter entitled "From What We Know to Faith: The Possibility of Revelation," Roberts rhetorically asked, "What is the meaning of the universe?" Listing around twenty-five statements and questions concerned with the "meaning of life," Roberts finally concluded that "in all that we have contemplated in our review of what humans know, we have found nothing that brings a solution to these inquiries." Citing Tom Paine's *Age of Reason*, Roberts argued that reason alone (here represented by Paine) is utterly inadequate in answering ultimate questions. Under the heading "Testimony of the Works of Nature Inadequate," Roberts argued:

> The universe itself conveys no information on these matters. "Turn not to that inverted bowl men call the sky," for answer to these questions; for the worlds of the universe are impotent to answer. I know how forceful in testimony the heavens and the glory of them can be in supplementing a certain positive message. . . . The heavens and the glory of them, however,

are and can be only auxiliary witnesses to the principal message that shall impart the knowledge we seek. Until that knowledge comes, however, appeal to the creation is vain in hope of finding anything conclusive upon the questions that are here presented.[24]

In the succeeding chapter, "Seekers after God: Revelation," Roberts concluded that only revelation could provide such final knowledge.

## James E. Talmage as Mormon Cosmologist

The oldest of the four apostles and the first to be called into the Quorum of the Twelve (in 1911) was James E. Talmage (1862–1933).[25] Born in England, educated at Brigham Young Academy, Lehigh University, and the Johns Hopkins University, Talmage was given a Ph.D. diploma (1896) from Wesleyan University (Illinois). As a professor of geology and later president of the University of Utah, Talmage lectured widely on virtually all scientific subjects, including, of course, geology, the life sciences, and astronomy. As primarily a geologist, however, Talmage did not engage astronomical topics often, except as his interests reflected on cosmogonical issues, such as the origins of the earth and of humans. He lectured extensively throughout the church and authored *Jesus the Christ* and *The Articles of Faith*, perhaps the most influential noncanonical writings since the time of Joseph Smith.

In this regard he became well known for his strong support of the old-earth idea (uniformitarianism) tempered by his rejection of the evolutionary thesis regarding humans, adopting what might sympathetically be called "uniformitarian creationism." Talmage felt comfortable with these views and remained committed to them through a long and influential career within the church. In the year in which he was called as apostle, Talmage wrote in his "Coming of Man":

> Thus it is that in the early events of creation the earth came at last to a condition fitted for the abiding place of the sons and daughters of God, and then man came forth upon the earth. But the beginning of man's mortal existence upon the earth was not the beginning of man. He had lived before even indeed as he shall live long after the earth has passed away and been replaced by a new earth and a new heaven. It has been said that the body of man is the result of evolutionary development from a lower order of beings. Be that as it may, the body of man today is in the very form and fashion of the spirit of man except indeed for disfigurement and deformities.[26]

Reaffirming the premortal, divine spiritual nature of humans, Talmage's views reflected the church's 1909 statement in suspending judgment on the evolutionary thesis in general, though he did reject the view of the development of human beings by chance alone.[27] Talmage had already done as much in 1895, however: "But God in His infinite wisdom saw the end from the beginning and adopted a plan that needed no changing; only modifications have been introduced to suit special purposes."[28] Later, during the 1930 pre-Adamic controversy between B. H. Roberts and Joseph Fielding Smith, Talmage, as an old-earth geologist, affirmed that "there were no such pre-Adamite races and that there was no death upon the earth prior to Adam's fall is likewise declared to be *no* doctrine of the Church."[29] From his 1890 paper "The Theory of Evolution," to his 1895 address "Education and Infidelity," to his 1911 essay "The Coming of Man," and finally to his 1931 pre-Adamite thesis, Talmage was forceful in shaping the views of many Mormons and church leaders.[30]

He frequently used these issues to express his views on the nature of science and its relationship to religion. As a scientist, but first as a Mormon, Talmage viewed "correct" science and revealed religion to be in close harmony, because the author of both is God. Thus, "What is the field of science," asked Talmage rhetorically? "Everything. Science is the discourse of nature and nature is the visible declaration of Divine Will. . . . There is naught so small, so vast that science takes no cognizance thereof. . . . Nature is the scientist's copy and truth his chief aim."[31] Elsewhere he wrote that "among our young people I consider scientific knowledge as second in importance only to that knowledge that pertains to the church and Kingdom of God. . . . Nature as we study it, is but the temple of the Almighty."[32]

Thus during his earlier years as a practicing scientist, Talmage engaged science with the enthusiasm of a devout follower. For example, not only were Mormons interested in the plurality-of-worlds idea but they were also interested in the precise method by which these numerous worlds and planets had been formed. Laplace, the French physicist who had invented the 'nebular hypothesis,' the prevailing theory of planetary formation during the late eighteenth and nineteenth centuries, proposed that the formation of stars and planets had resulted from the rotation, contraction, and condensation of the primeval solar material and gases.[33] As theologian and cosmologist, Orson Pratt vacillated considerably on the issue of the nebular hypothesis, mostly considering it insufficient but sometimes believing it reasonably adequate.[34] Most Mormons who were aware of this cosmogony endorsed it fully, however, suggesting that both Abraham and Moses were also in agreement. For example, "Moses understood the theory of nebular formation; Abraham also was versed in the laws of worldframing"; and "There is no real conflict between the Mosaic account of the creation and the nebular theory which is the most widely accepted among the most prominent scientists."[35] Joseph Smith himself had suggested that planets were organized from pre-existent material: "This earth was organized or formed out of other planets, which were broken up and remodeled and made into the one on which we live."[36]

Contrary to Joseph Smith's understanding, however, the nebular hypothesis presented a view that conflicted with the one he espoused. For a brief time during the first two decades of the twentieth century, the nebular hypothesis was replaced by the 'planetesimal hypothesis,' the view that the planets and satellites were formed as a result of a close encounter of the sun with another star.[37] Under this view, as our ancestral sun was approached by an intruder star, planets and satellites began to form from the accretion of parts of other pre-existent materials torn away from the sun and the star themselves. As one might expect, because the planetesimal hypothesis seemed to be vaguely similar to the ideas of Joseph Smith on planetary formation, Mormon scientists and authorities, among them Talmage, publicized this view broadly.[38]

Unfortunately, the planetesimal hypothesis requires extremely close cosmic encounters that are both empirically nonexistent and theoretically nearly impossible. Thus if the hypothesis were true, the likelihood of numerous stars possessing 'worlds without number,' to use Enoch's phrase, would be decreasingly unlikely. Consequently, this particular part of the hypothesis was simply ignored by Talmage. When the 'planetesimal hypothesis' lost scientific credibility around 1940, Mormons simply ignored the entire issue.

Although Talmage viewed scientific knowledge as nearly sacred, he recognized the essential importance and relative relationship between scientific facts and scientific theories. To be sure, "theories have their purpose and are indispensable in fact, but they must never be mistaken for demonstrated facts."[39] Still, "science is not merely knowledge; a simple accumulation of facts. . . . The parts [or facts] must be placed in proper relative position [as explained by theories], and only as this true relationship is established and maintained, will the structure approach completeness [or true science]."[40] Whereas these views were expressed mostly during the 1890s, Talmage continued throughout his long career as scientist and later church authority to maintain their validity. In short, although he reflected the realism sanctioned by the Mormon canon concerning an objective reality, he tempered that enthusiasm with a critical view of theory and fact.

## John A. Widtsoe as Rationalist

Born in Norway, John Andreas Widtsoe (1872–1952) was educated at Harvard (1894), taking his Ph.D. (1900) in chemistry at the university in Göttingen, Germany, one of the world's leading research institutions.[41] In Utah he was a professor at the Agricultural College in Logan (1894–98) and later at Brigham Young University (1905–7). He became president first at Logan (1907–16) then, before his call as an apostle in 1921, at the University of Utah (1916–21). He authored seven scientific books, scores of technical research papers, and numerous church-related manuals, books, pamphlets, and course materials. Throughout all his many years as scientist and religious authority, perhaps the most consistent theme in his work was the value of science for greater religious understanding. In this he espoused a rationalism in theology as a means of gaining increased understanding both of God's words and works and of the process needed to acquire that knowledge.

In his pre-apostolic book *Joseph Smith as Scientist* (1908), under a section entitled "The Fundamental Concepts of the Universe," Widtsoe discussed three "laws" of science that he claimed were considered basic scientific principles at the time: the indestructibility of matter, the conservation of energy, and the luminiferous ether. In all three cases Widtsoe asserted that Joseph Smith's views on these matters not only predated the scientific understanding of them but also that they were vindicated by science after their "discovery": "[I]t is not improbable that at some future time, when science shall have gained a wider view, the historian of the physical sciences will say that Joseph Smith, the clear-sighted, first stated correctly the fundamental physical doctrine of the universal ether [and the indestructibility of matter and of energy]."[42]

Strictly, neither matter nor energy alone is conserved; only together, in a combined system, will mass-energy be conserved. To this extent, therefore, matter and energy are indestructible. On the issue of the scientific ether, actually it was Orson Pratt, not Joseph Smith, who first equated it with a ubiquitous substance throughout the universe, namely the Holy Ghost (see chapter 6). Unfortunately, as early as 1905 Albert Einstein had already presented his theory of special relativity, in which he showed that these three assertions were either fundamentally misunderstood or simply incorrect. The ether, which eventually became a relic of nineteenth-century speculations that really extends back to Newton's views of the cosmos in the seventeenth century, had already first been called into question by Albert Michelson in 1881. It was not finally discarded, however, until Einstein's conclusions became gradually accepted nearly three decades later, around 1915.[43]

In *Joseph Smith as Scientist*, Widtsoe repeatedly argued that (1) Joseph Smith made specific scientific claims, (2) scientists subsequently and independently developed theories that conformed to these assertions,

John A. Widtsoe (1872–1952), one of Mormonism's first academically trained scientists and one of its most influential general authorities of the twentieth century. Courtesy of the Historical Department, Church of Jesus Christ of Latter-day Saints

# JOSEPH SMITH

AS

# SCIENTIST

A CONTRIBUTION TO
MORMON PHILOSOPHY

BY

John A. Widtsoe, A. M., Ph. D.

THE GENERAL BOARD
YOUNG MEN'S MUTUAL IMPROVEMENT
ASSOCIATIONS
SALT LAKE CITY, UTAH
1908

John A. Widtsoe's *Joseph Smith as Scientist* (1908). In this book Widtsoe described Joseph Smith as revealing many of the basic laws that later came to characterize modern science. Courtesy of the Special Collections Department, University of Utah Libraries and (3) because true science is revealed to prophets, Joseph Smith received his understanding directly from God. This is Widtsoe's argument, not Joseph Smith's, and if carried to its extreme would postulate a full-fledged natural theology, the results of which would be catastrophic. This was a problem for Widtsoe, however, not for Joseph Smith, because Joseph Smith never attached any ontological status to his claims. Therefore, it is more accurate to consider *Joseph Smith as Scientist* more as a reflection of the author's enthusiasm and his imaginative mind than as support for Joseph Smith's prophetic role.

Although these fundamental laws of nineteenth-century science were discredited by twentieth-century physics, Widtsoe, Roberts, and others asserted that other "scientific" ideas (not laws) were embraced by both Mormonism and the science of their day—particularly the plurality of worlds. Widtsoe noted that Joseph Smith taught that (1) stellar bodies are distributed throughout space, (2) stars evolve and eventually are destroyed, (3) the laws of astronomy are universal, (4) stars are in continual motion, (5) stellar bodies are organized in "great groups, controlled (under gravitational influence) by large suns," and (6) planets and stars throughout the celestial expanse are inhabited by sentient, rational beings.[44] Thus, concluded Widtsoe, "the doctrines of . . . [Joseph Smith] are in full accord with the views that distinguish the new astronomy." And finally, wrote Widtsoe rhetorically, "did Joseph Smith teach these truths by chance? or, did he receive inspiration from a higher power?"[45] As we have already illustrated, to the degree that Widtsoe was lost in his enthusiasm for science, he reflected nineteenth-century American attitudes toward science as a complete and finished enterprise.

During the 1920s, even after Widtsoe's call as an apostle in 1921, he continued strongly to espouse the virtues of his rational science. For example, in his 1927 YMMIA manual, *How Science Contributes to Religion*, Widtsoe emphasized the importance of the "scientific spirit":

> Science consists of the sound knowledge, the experience of mankind; it is, therefore, the truth possessed by humanity. The aim of science is to add knowledge to knowledge, so that the horizon of truth may be enlarged.
>
> The importance of science to mankind is partly in the revelation of new knowledge; but more in the development of man's love of truth. The love of truth that has been fostered by science is a profound power for good in our present-day world. Scientific knowledge or discovery is of little consequence in comparison with the quality of the force which drives the man into scientific activity.
>
> The scientific spirit requires at least four things of every disciple. First, he must love truth above all else. Second, he must, for himself, search out

truth, both old and new. . . . Third, he must be willing to hear and to test all claims for truth, that may present themselves, and to sift such claims on the basis of naked evidence irrespective of his existing prejudices. Fourth, he must be so openminded that when truth appears he will recognize it and accept it, though it is contrary to all former opinions.[46]

Continuing this theme in his book *In Search of Truth*, published in 1930 for the general church body, Widtsoe addressed the issues in chapters entitled "What Is Science?" and "How Does the Church View Science?" His advocacy for a rational, positive scientism is undeniable:

Science . . . is the recognition by the mind through human senses of the realities of existence. The mind of man is a noble instrument, a pre-eminent possession, by which he becomes conscious, not only of his own existence, but of the conditions of external nature. By tests of mind, humanity is distinguished from the lower orders of life. The triumphs of science are evidences of the supremacy of mind—a supremacy which dominates all nature. The glory of physical conquests, of the sea and earth and air, have often dazzled men to such a degree that they have forgotten that back of all discovery and progress is the power of observation and thought. Without mind, there is no science, no progress, only extinction.[47]

Through most of his many years as scientist and later church leader, Widtsoe continued to promote science and scientific thinking in the strongest possible terms. Yet, as we shall see in the last section, the years immediately following 1930 represent a sort of watershed in his thinking, and, as with Roberts and Talmage before him, Widtsoe eventually tempered his scientific excessiveness from the 1930s onward.

### Joseph F. Merrill as Science-Faith Advocate

Joseph F. Merrill (1868–1952) was educated at the University of Michigan (1893) and completed his Ph.D. in physics at the Johns Hopkins University (1899), America's first research university patterned after the German system. Although Merrill was not as forceful an advocate of scientism as perhaps Widtsoe or Talmage or even Roberts, nevertheless he frequently spoke and wrote on scientific topics, particularly as they related to enhancing the religious faith among Latterday Saint youth. During his academic career as professor of physics, chemistry, and engineering at the University of Utah (1893–1928), he engaged in science as a teacher and scholar, rarely using his science to explore wider religious issues. Only after he left academia to serve as church commissioner of education, accepting various ecclesiastical positions including apostle in 1931, did he publicly address questions dealing specifically with the relationship of science and religion.

Merrill's scientific philosophy was fully in keeping with that of Widt-soe, Talmage, and most other scientists, Mormon and otherwise, at the time. Thus he adopted the view of the primacy of natural law as an operative principle in the organizing of the cosmos and life within it.[48] He affirmed the view that a natural causation underlies this cosmos and that through this orderliness of the universe science is able to reflect a true knowledge of God.[49] Because in Merrill's mind this cause-and-effect assumption shows that God is ultimately the author of nature, he denied that natural causation implies that the universe evolved from chance.[50]

Like his scientific colleagues within the Quorum of the Twelve, Merrill espoused a deep commitment to a teleological view of the cosmos. Thus, for instance, not only was eternal progression possible in spiritual matters, but the physical universe itself is in a state of constant development and progression.[51] Even science, as well as human society in general, is becoming increasingly enlightened. Thus in Merrill's view, although nineteenth-century science was characterized by a materialist creed that ruled out the presence of God, around 1900 that philosophical materialism was slowly being discarded as the result of the influence of two scientific events. First was the realization that science as a collection of discoveries and facts was not complete, that there were additional discoveries to make. And second was the discovery of radioactivity, which not only opened up a whole new dimension of scientific research but, more importantly, demonstrated that nature (and hence God) is much more complex and deep than hitherto understood.[52]

On the nature of scientific ontology, Merrill vacillated. On the one hand, he believed that scientific knowledge (as with religious understanding) comes after much reflection. Such inspiration or intuition implies, however, that scientific ideas have a direct correspondence to reality.[53] Even though this view is supported by statements in the Mormon canon, as we have seen and as Merrill recognized, science changes too much to sustain this interpretation. On the other hand, Merrill's enthusiasm for science was partially tempered by Widtsoe's excessive tendency to wed science with truth. Merrill did speak, for instance, of the plurality-of-worlds doctrine in science, but he also recognized that anachronistic scientific ideas, such as the ether concept, compelled a certain scientific caution.[54] In the final analysis, Merrill adopted the view that science cannot know ultimate reality, and therefore the scientist must also walk by faith. Neither the methods nor the findings of science, argued Merrill, can *prove* the existence of God. At best they will reveal His works and hence we may infer His existence.[55]

This mild scientific caution urged Merrill from the excesses of a natural theology.

Merrill continued to be a strong science advocate, particularly in religious matters where science could reveal a purposeful universe. Thus despite his reluctance to assent to the view of the ultimate truthfulness of science, he followed the long Mormon tradition and firmly believed that science, rightly understood, does not contradict religion.[56]

## Mormonism's Retreat from Realism

Those Mormon scientists who now functioned in the church hierarchy continued to advocate a rational scientism. Whether Widtsoe's views on the conservation of matter and energy as well as on the ether, or Talmage's ideas about uniformitarian creationism, or Roberts's advocacy of pre-Adamites and the nature of extraterrestrial life, however, this confidence brought them dangerously close to a full-fledged natural theology. Trained in the rarefied atmosphere of scientific research by some of the best graduate schools of science in the world, Widtsoe, Talmage, Merrill, and Lyman, as well as Roberts, used the full force of their scholarship in the service of harmonizing science with their religion. Mormonism was truth revealed, and science was truth emerging. Although they mostly understood the dangers in a natural theology, in the final analysis their religion tempered excessive enthusiasm for a science-based theology. But whereas science, as natural theology, was somewhat cautioned as a discipline, it was nevertheless advocated in a positive and constructive dimension that would continue attitudes that flowed from a Mormonism that viewed the world as God's creation.

In the years preceding the Second War, however, not only did Widtsoe and Merrill begin to express reservations about an uncritical blending of science with theology, but as they began to view science as technology—as application—they became somewhat less enthusiastic advocates for science. This is not to suggest that they replaced their rational scientism with an irrationalism; rather, they began to argue that, as a *human* enterprise, science is fundamentally a tentative and opaque study justified ultimately in terms of its applications to humankind. Their views, however, reflected larger trends within the American intellectual community. As we have already noted, most Americans throughout the nineteenth and early twentieth centuries had confused science with technology and perceived technology as providing an increasingly greater material abundance. This scientific-technological attitude was largely responsible for reinforcing Americans' views of progress. In the aftermath of the First War and the alleged decline of Western civilization in the early decades of the twentieth century, this attitude began to wane considerably.[57] Moreover, the scientist of the nineteenth century was characterized by a refreshing, though somewhat misleading, amateurism. Still, he was a dilettante in quest of understanding; he was not yet, as Charlie Chaplin portrayed in *Hard Times*, caught in the wheels of big business and industry, to be exploited for profit and greed. The romance of "little science," with modestly equipped laboratories in the nineteenth century, was replaced in the twentieth century by a "big science" driven by huge, complex, and expensive machines increasingly funded by large amounts of private, and eventually public, capital. Science thus joined big business and was increasingly used as a key component in the rise of the military-industrial complex.[58]

Although these trends were primarily social and technological, there were also some profound changes philosophically. If the seventeenth century laid the foundations for modern science, it was the fulfillment of the Newtonian worldview in the nineteenth century, which, it was widely believed, would culminate in a complete knowledge of nature and her workings. As the twentieth century dawned, however, a crisis of faith within the scientific community fundamentally and permanently ruptured this complacency. With the emergence of relativity and quantum mechanics, not only did physics reject its Newtonian foundation but the discovery of radioactivity and a subatomic world urged in deep ways a rejection of traditional ideas about matter itself. During the 1920s the astronomer's views of the cosmos began to shift radically from eighteenth- and nineteenth-century ideas. The world of science forever afterwards was fundamentally different than the one before 1900. And finally, and perhaps most important scientifically, the cause-and-effect assumption of the Newtonian world had been replaced by a probabilistic approach to the universe. No longer was nature seen as being completely transparent to human understanding.

Ultimately, all these church authorities came to view science not as an end in itself but rather in the service of humankind. In this sense science is not philosophy but technology and application. During their years directing the church's missionary work continuously in Europe from 1924 to 1927, 1927 to 1933, and 1933 to 1936, respectively, Talmage, Widtsoe, and Merrill saw first hand the effects of a technological scientism. The rise of fascism and communism was a dangerous portent as Europe increasingly struggled with intense moral and social problems. Here Talmage, Widtsoe, and Merrill put aside their science and adopted the role of ecclesiastical leaders for a church that itself was still struggling for respect among its religious neighbors. Europe seemed bankrupt; the forces of good and evil were arrayed against one another. As a result, their experiences in Europe at this time shifted from the abstract philosophizing of science to the more mundane, immediate press of a failed European order. Though not published until 1945, Widtsoe's *Man and the Dragon,* a collection of editorials written for the *Millennial Star* during the years he served as European and British mission president, was based on the premise of "man's constant but victorious battle with the forces of evil [that issue] from 'the dragon, that old serpent, which is the devil, and Satan.' "⁵⁹ Since Roberts and Talmage would both die in 1933 and Lyman was soon to lose all influence, it was Merrill and particularly Widtsoe who, in the years immediately preceding the Second War, most strongly reflected Mormonism's temporary retreat from an *applied* scientism.⁶⁰

Neither Widtsoe nor Merrill, however, relinquished his promotion of science as a positive, rational endeavor that provides a greater understanding of the world about us. Addressing the attitude of the church toward science in his 1939 *Evidences and Reconciliations,* Widtsoe argued, in keeping with Joseph Smith, Brigham Young, Joseph F. Smith, and many others, that "the Church supports and welcomes the growth of science. . . . The religion of the Latter-day Saints is not hostile to any truth, nor to scientific search for truth."⁶¹ Although he continued to maintain his enthusiasm for science as philosophy, after about 1930 his experiences, particularly in Europe, urged him also to acknowledge what both Roberts and Talmage already recognized—that science has limits that inherently expose its potential failings. For the first time he was willing to argue that, because science is in a constant state of flux, scientific hypotheses and theories in one era may be discarded in another. Ultimately, wrote Widtsoe in his *Evidences and Reconciliations,* "science is trustworthy as far as human senses and reason are trustworthy—no more. When the credentials of science are examined, the claims of religion seem more credible than ever."⁶² Whereas Merrill continued to promote science as utility, he was not far behind in urging the epistemological caution expressed by his fellow apostle Widtsoe.⁶³

## NOTES

1. Their dates of service as general authorities are: B. H. Roberts, 1877–1933; James E. Talmage, 1911–33; Richard R. Lyman, 1918–43; John A. Widtsoe, 1921–52; and Joseph F. Merrill, 1931–52. Although Lyman became an apostle in 1918, his influence declined during the 1930s and ceased altogether after his excommunication in 1943.

2. See, for example, Thomas Alexander, "Reconstruction of Mormon Theology: From Joseph Smith to Progressive Theology," *Sunstone* 5, no. 4 (July/August 1980), 24–33.

3. See, for example, Leo Marx, *The Machine in the Garden: Technology and the Pastoral Ideal in America* (New York: Oxford University Press, 1967).

4. Joseph Fielding Smith was the strongest antagonist of science among the most influential leaders of the church. His views, particularly on evolution, appeared first during the 1920s and were in direct and heated opposition to those of B. H. Roberts. For a full discussion, see chap. 8.

5. Davis Bitton, "Anti-Intellectualism in Mormon History," *Dialogue* 1, no. 3 (Autumn 1966): 111–34, esp. 119–23; and Leonard Arrington, "The Intellectual Tradition of the Latter-day Saints," ibid. 4, no. 1 (Spring 1969): 13–26.

6. John A. Widtsoe, *Joseph Smith as Scientist* (Salt Lake City: YMMIA, 1908), 9, first published in a series of articles in *Improvement Era*, 1903/4.

7. See Truman G. Madsen, *B. H. Roberts: Defender of the Faith* (Salt Lake City: Bookcraft, 1980).

8. See Arrington, "The Intellectual Tradition of the Latter-day Saints," 24; and Sterling M. McMurrin, "Brigham H. Roberts: Notes on a Mormon Philosopher-Historian," *Dialogue* 2 (Winter 1967): 141–49, reprinted from McMurrin's introduction to B. H. Roberts, *Joseph Smith: The Prophet-Teacher* (Princeton: Deseret Club of Princeton University, 1967).

9. Truman G. Madsen, "The Meaning of Christ—'The Truth, the Way, the Life': An Analysis of B. H. Roberts's Unpublished Masterwork," *Brigham Young University Studies* 15, no. 3 (Spring 1975), 259–92.

10. For the details of this conflict, see chap. 8.

11. Frederick J. Pack, *Science and Belief in God* (Salt Lake City: Deseret News, 1924), and Nels L. Nelson, *Scientific Aspects of Mormonism* (New York: Dutton, 1904).

12. B. H. Roberts, "The Truth, the Way, the Life: An Elementary Treatise on Theology" (1928), chap. 3, 1.

13. Ibid.

14. Roberts, "The Truth, the Way, the Life," chap. 3, 11–15. See Robert Kargon, *The Rise of Robert Millikan: Portrait of a Life in American Science* (Ithaca: Cornell University Press, 1982).

15. The Copernican Revolution of the sixteenth century was the first astronomical revolution. On the emergence of the "new" astronomy, see Richard Berendzen, Richard Hart, and Daniel Seeley, *Man Discovers the Galaxies*, (New York: Science History, 1976); Robert Smith, *The Expanding Universe: Astronomy's "Great Debate," 1900–1931* (New York: Cambridge University Press, 1982); and Erich Robert Paul, *The Milky Way Galaxy and Statistical Cosmology, 1890–1924* (forthcoming from Cambridge University Press). Shapley's estimates for the size of the Milky Way system were later reduced by a factor of two-thirds (to 100,000 light years diameter) to conform to additional empirical evidence.

16. F. R. Moulton, quoted in Roberts, "The Truth, the Way, the Life," chap. 5, 5.

17. Roberts, "The Truth, the Way, the Life," chap. 6, 1.

18. Ibid., chap. 7, 4. By contemporary standards Roberts's statement is not unproblematic, because present-day twentieth-century astrophysics now asserts that the universe began originally in a big bang, prior to which there was simply nothing (at least we can never know of anything prior to this singular event). For some intriguing religious implications of this view, see Robert Jastrow, *God and the Astronomers* (New York: Warner Books, 1978), and Keith E. Norman, "Mormon Cosmology: Can It Survive the Big Bang?" *Sunstone* 10, no. 9 (1985), 19–23.

19. Roberts, "The Truth, the Way, the Life," chap. 7, 5.

20. Ibid., chap. 10, 1–2.

21. Ibid., chap. 10, 9.

22. Ibid., chap. 10, 12.

23. Ibid., chap. 10, 14–15. Roberts quoted the relevant sections of Abraham, chap. 3 and 4, and D&C 76:37–39, 45.

24. Ibid., chap. 11, 2–3.

25. For biographical details of Talmage's life, see John R. Talmage, *The Talmage Story: Life of James E. Talmage—Educator, Scientist, Apostle* (Salt Lake City: Bookcraft, 1972); and D. Rowley, "Inner Dialogue: James Talmage's Choice of Science as a Career, 1876–1884," *Dialogue* 17, no. 2 (Summer 1984), 112–30.

26. James E. Talmage, "Notes—The Coming of Man," Ms. 229, Box 22, Folder 8, Brigham Young University Special Collections, Harold B. Lee Library.

27. For a discussion of the evolution issue in Mormonism, see chap. 8.

28. James E. Talmage, "Education and Infidelity," (30 June 1895), Ms. 229, Box 22, Folder 7, Brigham Young University Special Collections.

29. James E. Talmage, personal journal, 7 April 1931, Brigham Young University Special Collections (italics added). For details of this debate, see chap. 8.

30. For a discussion of Talmage's role in the Roberts-Smith controversy, as well as related issues, see Richard Sherlock, " 'We Can See No Advantage to a Continuation of the Discussion': The Roberts/Smith/Talmage Affair," *Dialogue* 13, no. 3 (Fall 1980), 63–78; and Jeffery E. Keller, "Discussion Continued: The Sequel to the Roberts/Smith/Talmage Affair," ibid., 15, no. 1 (Spring 1982), 79–98.

31. James E. Talmage, "Science and Art," Ms. 229, Box 1, Folder 7, Brigham Young University Special Collections.

32. James E. Talmage, "Science in the Associations," Ms. 229, Box 22, Folder 7, Brigham Young University Special Collections.

33. On the 'nebular hypothesis,' see Ronald L. Numbers, *Creation by Natural Law: Laplace's Nebular Hypothesis in American Thought* (Seattle: University of Washington Press, 1977).

34. Orson Pratt, "1871 Lectures" (see England, *The Life and Thought of Orson Pratt*, 251); Pratt, "New Theory of the Origin of the Solar System," *Deseret News* 20 (15 November 1871), 476; and idem, *Journal of Discourses* 16:316

35. "Science and Genesis," *Deseret News* 29 (1880), 758; and "Science and Religion," ibid. 28 (1879), 166.

36. Joseph Smith, quoted in Frederick J. Pack, "Natural and Supernatural," *Improvement Era* 11 (January 1908), 182.

37. See Stephen G. Brush, "A Geologist among Astronomers: The Rise and Fall of the Chamberlin-Moulton Cosmology," *Journal for the History of Astronomy* 9 (February/June 1978), 1–41, 77–104.

38. James E. Talmage, "Prophecy as the Forerunner of Science—An Instance," *Improvement Era* 7 (May 1904), 481–88.

39. James E. Talmage, "The Coming of Man," Ms. 229, Box 22, Folder 8, Brigham Young University Special Collections.

40. James E. Talmage, "The Methods and Motives of Science," *Improvement Era* 3 (February 1900), 251.

41. For biographical details of Widtsoe's life, see Dale C. LeCheminant, "John A Widtsoe: Rational Apologist" (Ph.D. dissertation, University of Utah, 1977).

42. Widtsoe, *Joseph Smith as Scientist*, 29, chap. 1–4 passim.

43. The history of ideas on matter, energy, and the ether is long and complex. For reliable discussions of these issues and their eventual demise, see G. N. Cantor and M. J. S. Hodge, *Conceptions of Ether: Studies in the History of Ether Theories, 1740–1900* (New York: Cambridge University Press, 1981), and P. M. Harman, *Energy, Force and Matter: The Conceptual Development of Nineteenth-Century Physics* (New York: Cambridge University Press, 1982). Einstein's idea of special relativity was not motivated by the null effects of the 1881 and 1887 Michelson-Morley experiments; see Gerald Holton, "Einstein, Michelson, and the 'Crucial' Experiment," *ISIS* 60, no. 2 (Summer 1969), 133–97. Also, as with most scientific ideas, except for German physicists the acceptance of special relatively required nearly two decades; see T. F. Glick, ed., *The Comparative Reception of Relativity*, Boston Studies in the Philosophy of Science (Dortrecht: Reidel, 1987), 103.

44. Widtsoe, *Joseph Smith as Scientist*, 46–49.

45. Ibid., 49.

46. John A. Widtsoe, *How Science Contributes to Religion* (Salt Lake City, YMMIA, 1927), 7–8.

47. John A. Widtsoe, *In Search of Truth* (Salt Lake City: Deseret Book, 1930), 36–7.

48. Joseph F. Merrill, "A Peek into the Heavens," in *The Truth-Seeker and Mormonism* (Salt Lake City: Deseret Book, 1946), 12; and Merrill, "Our Subatomic Realm," in *Science and Your Faith in God*, comp. Paul Green (Salt Lake City: Bookcraft, 1958), 108.

49. Merrill, "Marvels Revealed by Science," in *Truth-Seeker*, 16–17; and idem, "Mysteries of Science and Religion," in *Science and Your Faith in God*, 126.

50. Merrill, "Marvels Revealed by Science," in *Truth-Seeker*, 20; and idem, "The Dynamic God of Science," in *Science and Your Faith in God*, 118–19.

51. Ibid., 119.

52. Merrill, "Does Science Prove the Existence of God?" 129–30.

53. Merrill, "Faith in the Realm of Science," in *Science and Your Faith in God*, 144–45.

54. Merrill, "A Peek into the Heavens," in *Truth-Seeker*, 13–14; and idem, "Electricity, Light and Materialism," in *Truth-Seeker*, 39, 42–43.

55. Ibid., 39; Merrill, "A Way to Discover God," in *Truth-Seeker*, passim; idem, "Is There a God?" in *Science and Your Faith in God*, 133.

56. Merrill, "Marvels Revealed by Science," 18, 25.

57. The literature documenting this view is extensive; for some period observations see, for example, Oswald Spengler, *The Decline of the West* (New York: Alfred Knopf, 1929); and Stefan Zweig, *The World of Yesterday* (Lincoln: University of Nebraska Press, 1971). For a provocative examination of some philosophical, cultural, and artistic implications of the nineteenth-century background, see Stephen G. Brush, *The Temperature of History: Phases of Science and Culture in the Nineteenth Century* (New York: B. Franklin, 1978).

58. Derek de S. Price, *Little Science–Big Science* (New York: Columbia University Press, 1963; rev. ed., 1986).

59. John A. Widtsoe, *Man and the Dragon* (Salt Lake City: Bookcraft, 1945), 14.

60. For a discussion of Widtsoe's scientific ambivalence, see Le-Cheminant, *John A. Widtsoe: Rational Apologist*, chaps. 4 and 5. LeCheminant implies incorrectly, however, that Widtsoe retreated from science after he became an apostle (see 145–53).

61. John A. Widtsoe, *Evidences and Reconciliations*, 3 vols. (Salt Lake City: Bookcraft, 1943), 1:129.

62. Ibid., 133.

63. Merrill, *Truth-Seeker*, passim.

# 8

## Theology or Science:
## A Warfare of Ideas?

Although cosmology has been woven deeply into the theological fabric of Mormonism, the sciences of cosmogony and evolution have been less well received. Principally this is because the latter deal with *origins*—the origin of the Universe and the origin of humans and life in general, respectively. And because questions of origins eventually concern themselves with beginnings, the role of God in the physical world is nearly always paramount. Thus the sciences of origins, such as cosmogony and areas of evolution and geology, treat issues frequently addressed simultaneously in sacred scripture.

In modern cosmogony, the big-bang theory of the beginning of the universe provides a powerful understanding that may be taken to verify the story in Genesis concerning origins. Indeed, this confluence of traditions has not gone unnoticed; some have argued that modern scientific theories about the beginning of the physical universe coincide with the story that the theologians have been telling us all along.[1] The big-bang theory deals fundamentally with *origins*, however—origins approximately 15 billion years ago. Consequently, to adopt this view of the universe compels questions about the creation and development (evolution) of both the physical and the organic, about geology and life.

Although the cosmogony of the big bang may be seen to support, however loosely, the Genesis story, in the Mormon tradition an antiscientism, particularly toward the sciences of geology and evolution, has nevertheless developed.[2] Following the post-Darwinian controversies around the turn of the century, there emerged within some quarters of Mormonism the strident view that modern science (or at least certain parts of it) is antithetical to the restored gospel.[3] In more recent decades this has been supported by the persistence of a theological literalism and a drift to the ideological right that has deeply affected the relationship between science and Mormonism.[4] Ironically, some have actually gone so far as to provide a pseudoscientific basis to their theology to bolster their position. They have done so by providing a scientific veneer for some of their ideas—a veneer which (1) has been totally rejected by the modern scientific community as too highly speculative, (2) is based on views that are not strictly scientific but more religious, and (3) has become scientifically outdated and therefore no longer tenable. Such views have resulted in the desire among some to institutionalize a reactionary science, one in which mainstream science plays a distinctly inferior role. Using various church avenues, some insist on indoctrinating the Mormon body politic—without full disclosure—with the view that an outmoded (nineteenth-century) science and cosmology is conceptually true.

Although this reactionary view would be understandable in an earlier and less sophisticated era, what makes the more recent emergence of this antiscientism particularly distressing is the fact that it comes in an age when science has conclusively rejected reactionary speculation. Underpinning this neoliteralist interpretation among theologically conservative Mormons is the philosophical view that one can know the workings of nature (reality) by engaging merely in biblical exegesis of those accounts of scripture allegedly dealing with natural phenomena, such as the creation of the world as given in Genesis, the geocentric astronomy of the Old Testament, as well as other now discredited scientific ideas found within various scriptural sources.

In recent decades Mormonism as a body has grown enormously. This has complicated matters for those who adopt reactionary views, primarily because the Mormon church now has large numbers of highly educated scientists and scholars who know that this kind of thinking is no longer tenable. Consequently, certain tensions have arisen within the institutional church that have begun to pit one group of intellectuals against another group of self-professed theologians.

## Cosmogony and Mormon Creation Narratives

The Condemnation of 1277 and the rise of empirical science in the seventeenth century are both related closely to the belief in creation ex nihilo, a doctrine accepted by Christians in the second and third centuries owing to their belief in the literal resurrection of Christ. Biblical scholarship has shown, however, that the creation process as understood within traditional Christianity may have nothing to do with the ex nihilo Christian doctrine. Indeed, scholars have noted not only that there are two creation narratives in Genesis (Genesis 1–2:4a and Genesis 2:4b–4:27)—neither of which may have anything to do with an actual "scientific" creation of the world in six days (or periods), let alone an ex nihilo creation—but that the ex nihilo teaching is almost certainly postbiblical.[5]

On this issue Mormons in the literal-historic tradition and biblical scholars agree, but for vastly different reasons. Neoliteralist Mormons deny creation ex nihilo but ascribe a literal sequence to the creation accounts as providing an actual physical—scientific—explanation of the beginnings of the world.[6] The only contemporary biblical tradition that still adopts simultaneously creation ex nihilo and Genesis as describing a physical event is that held by ideologically conservative evangelical Christians.

In sharp contrast to both this evangelical tradition and neoliteralist Mormonism, biblical scholarship has largely rejected altogether a literal rendering of Genesis. Thus, for example, some have indicated that the first creation account in Genesis deals fundamentally with a theological problem—namely, monotheism vis-à-vis polytheism, syncretism, and idolatry. In this view, all of the non-Israelite cultures at the time that Genesis was written were polytheistic. As one scholar has described it, "each day of creation takes on two principal categories of divinity in the pantheons of the day, and declares that these are not gods at all, but creatures—creations of the one true God who is the only one, without a second or third."[7] Specifically, "light" and "darkness" (day one) are populated by the "greater light" and the "lesser light" (day four); "firmament" and "waters" (day two) are populated by "birds" and "fishes" (day five); "earth" and "vegetation" (day three) are populated by "animals" and "humans" (day six). Each of these categories represented a godly presence within the non-Israelite cultures of Egypt, Assyria, and Babylon. The purpose of Genesis, therefore, is to argue that these gods are not divine; rather, that the Israelite god—and the only *one* god—*created* each of these basic categories.

Although conceding that Genesis may ultimately deal with a quasi-creation of the world as understood in literalist traditions, some have argued that, in so doing, this account has drawn upon the mythic cultural background of the ancient Near East in order primarily to focus again on demythologizing the pantheon of pagan gods.[8] Still others have emphasized that, whatever the theological issue, the mythic account in Genesis has been deeply enriched by the two creation narratives. The focus of these accounts was to disassociate Israelite monotheistic religion from the mythical nature cycles advanced by her neighbors. As one Mormon scholar has described it: "Creation, as it functions in the Old Testament, is not a timeless, mystical drama that must be repeated periodically to ensure fertility or avert the wrath of the gods. It is, rather, the prelude to history and establishes the basis of humanity's relationship to God."[9] Freed thus from polytheism, creation becomes the starting point of history and biblical faith thus became anthropocentric.[10]

Such mythic accounts of the creation narrative stand largely outside the Mormon tradition, which in Mormonism's formative years was dominated by an unquestioning biblical literalism.[11] However, for two reasons they do not rule out the understanding of the various creation narratives found in Mormonism's unique canon. First, whether dealing with origins of the world in the scientific sense or with a mythic account of humankind's place in the grand scheme of things, creation in both biblical and Mormon traditions is not only directly related to the beginning of history but pre-eminently a religious affirmation of the sovereignty of God and the total dependence of humans on their Creator.[12] Second, the idea of a prophet, in both biblical and Mormon views, is one who, although possessing the possibility of foretelling future events, has the mandate to interpret and reinterpret all creation accounts.[13] Whereas traditional accounts of the creation lead us to understand that Genesis may deal primarily with a theological issue, additional and complementary interpretations are both allowed and expected.

Thus, as we have noted in chapter 5, during Joseph Smith's most mature period he introduced yet another creation narrative. Enoch (in Moses 2 and 3) outlines a spiritual creation of the heavens and the earth preceding the physical creation as given by Abraham (in Abraham 4 and 5). Thus Enoch explains that God "created all things spiritually, before they were naturally upon the face of the earth" (Moses 3:5), affirming the spiritual nature of the cosmos. In this view God created all things—including his offspring as well as the cosmos itself—spiritually first; and only afterwards He patterned His creation physically.[14]

Precisely because Mormonism lacks a rigid, doctrinaire theology and because its tradition had already adopted from earliest times this dual account of the creation, Mormonism is broad enough to encompass mythic accounts. As with orthodox accounts of the creation dealing fundamentally with a theological problem, Mormonism too has interpreted the creation fundamentally as a theological issue—namely, the affirmation of the spiritual relationship of God to humans. It is this feature of Mormon scripture that ultimately must reject attempts to harmonize the first account of the creation exclusively with the scientific version of creation. Although neoliteralism is still strongly prevalent, given its denial of creation ex nihilo and its assertion of a spiritual creation, twentieth-century Mormonism has theologically freed itself from the demands of its earlier literalist roots.

## Mormonism and Evolution

By the last years of the nineteenth century, some scientists were suggesting that on two points Darwinian evolution might be problematic. Evolution by means of natural selection demands the continuance of some hereditary material, something to pass on to subsequent generations the accumulated characteristics of the ancestral stock. Darwin himself did not know what this hereditary material might be. Because the laws of Mendelian genetics, discovered in the 1860s, did not become generally known until 1900, nearly two decades after Darwin's death in 1882, for nearly four decades—1859 to 1900—evolutionary scientists could not deal with the critical question of heredity.[15]

Of perhaps greater impact was the time scale needed for the natural evolution of species. During the last decades of the nineteenth century mathematical physicists had calculated that there was at most enough energy in the sun (thought to have an incandescent liquid core, later changed to a dense gaseous material) to last between 10 million and 100 million years. Indeed, the amount of material in the sun was vastly inadequate to sustain long-term heat, so some scientists advanced the view that the burning sun was constantly replenished by comets spiraling onto its surface. Evolution seemingly required hundreds and thousands of millions of years. Among those demanding adherence to this view was Lord Kelvin, one of the most widely respected scientists of the time. Kelvin and many others argued that, given the rapid burning of the sun's source of fuel, there simply was not enough time for development of the multitudinous species that in fact do exist. Both physical scientists and orthodox Christians used the mathematical calculations and theories of physics to discredit evolutionary views.[16] Thus until the very end of the nineteenth century, it appeared that on both these counts—heredity and limited geological time—an apparent conflict between evolution and creationism could be resolved only with the total rejection of the Darwinian position.

But this did not happen. Around the turn of the twentieth century, the discovery of radioactivity suggested a new source of solar energy that would allow virtually unlimited time for the evolution of the nat-

ural universe. This quelled Kelvin's objections. In addition, the rediscovery of Mendelian genetics in 1900 saved Darwin's theory of evolution from virtual collapse. Thus Mormons and others wishing to
maintain a traditional reading of scripture could not simply reject the
evolutionary and biological arguments out of hand: evolution had to
be dealt with on its own terms as a possible scientific explanation for
the origin and subsequent development of all living organisms and
systems.

Mormons saw science, in concert with natural theology, as a means
to understand the works of God. Because most scientists were also
Christian and most Christians were interested in science with respect
to natural theology, questions about the creation of the world and the
origin of life and humans were crucial. Although by the early years of
the nineteenth century a few scientists and philosophers were beginning to dispute the literal interpretation of the biblical chronology, the
generally received Christian view was that the earth had been created
ex nihilo beginning the night preceding 23 October 4004 B.C.[17] Humankind and all living things were similarly created as the physical
embodiment of some Platonic idea *(eidos)* in the mind of Deity.[18]

During the last half of the nineteenth century, the increasing dominance of Darwinian thinking gradually transformed Christianity and
the cultural climate in America.[19] In keeping with most Christians of
the period, the vast majority of Mormons continued to resonate to a
traditional view of the origin of humans and life: man's body particularly was the direct, active, and immediate creation of God. In the
decades prior to Darwin science would have supported this creationist
view of humans and life; after Darwin, however, science increasingly
and persuasively argued for a naturalistic alternative. Those Christians
caught in the web of natural theology, trying to substantiate their religious claims by recourse to the findings and theories of science, often
lost their faith with the changing developments in science. However,
because of their insistence on "divine" and "inspired" science, Mormons who remained biblical traditionalists on this issue did not feel
threatened by geological and evolutionary studies. True science would
not contradict revealed religion—that is, religion that is prophetically
given. Whereas Christians had no final arbiter, Mormons had their
prophet.

In Mormon publications references to the origin of man were as
prominent in the late nineteenth century and early years of this century
as they are today. Because speculation on the topic of origins increased
after Darwin's discoveries, the First Presidency of Joseph F. Smith issued a statement in 1909, on the centennial of Darwin's birth, "of the

position held by the Church upon this important subject." Noting that (1) humans were created in the image of God and (2) the creation was two-fold, first spiritual and second temporal, both assertions having been given repeatedly by divine inspiration and therefore beyond dispute, they asked, rhetorically, "What was the form of man, in the spirit and in the body, as originally given?" Answer: "It is held by some that Adam was not the first man upon this earth, and that the original human being was a development from lower orders of the animal creation. These, however, are the theories of men. . . . [A]ll men were created in the *beginning* after the image of God; and whether we take this to mean the spirit or the body, or both, it commits us to the same conclusion: Man began life as a human being, in likeness of our heavenly Father."[20]

Although some considered this statement an endorsement of antievolution, for those either inclined or opposed to evolution, the First Presidency's 1909 statement discreetly avoided comment on the actual modus operandi of the creation process itself. In the April 1910 issue of the *Improvement Era*, Smith issued the following clarifications: "Whether the mortal bodies of man evolved in natural processes to present perfection, thru the direction and power of God; whether the first parents of our generations, Adam and Eve, were transplanted from another sphere, with immortal tabernacles, which became corrupted thru sin and the partaking of natural foods, in the process of time; whether they were born here in mortality, as other mortals have been, are questions not fully answered in the revealed word of God."[21] In 1925 the First Presidency of Heber J. Grant reaffirmed this view by reissuing the 1909 statement on "the origins of man"—with the antievolution references above conspicuously omitted.[22] In short, although the church has unequivocally affirmed the spirit heritage of human beings, it has disclaimed absolute knowledge of the actual process of creation of the physical body.

Given all these First Presidency messages, and in contrast to many Christians, Mormons had the option of remaining relatively free of any sort of scientific-theological dilemma. Theologically conservative Mormons, however, have ignored the 1910 and 1925 statements and considered the 1909 statement as documentation in support of an inflexible anti-evolution posture. Others regarded the 1909 statement in its *entirety* as highly ambiguous. With three-quarters of a century of hindsight today, the 1909 statement is at best misleading; at worst, one of the sources of the antagonistic stance toward evolution in particular and science in general that pervades some thinking among Mormons today. Like mainline Christians, deeply divided on the issue of evolution, becoming either biblical literalists and increasingly fundamental in their theological understanding or increasingly secular and even antireligious, Mormons also suffered a deep division. On the one hand, most Mormons consistently maintained a positive attitude toward science, claiming that God operates using natural laws and that, in principle, there is no inherent conflict between true science and revealed religion. On the other hand, some, increasingly threatened by evolutionary thinking, adopted a rigid biblical literalism.[23]

Mormonism, however, is theologically in the strongest possible position vis-à-vis biblical literalism. A central dogma of the Mormon faith holds that humans stand in a *necessary*, versus contingent, relationship to God. That is, man and God are co-extensive ontologically, each requiring the other in fundamentally essential ways. Furthermore, argue Mormons, God eventually propagates spiritually embodied beings in some sort of literal way. Thus humans are both co-extensive with, and filial to, God. The essential bonding between humans and God is then as child to parent—His children are His literal spiritual offspring. After God created the physical universe to provide (among other things) a physical residence for His spiritual offspring, humans became corporeal—in a *physical* body.

For Mormons the essential defining nature of humans, therefore, is not the physical body but the view that humans are the literal spiritual *offspring* of the creator. Whether the physical body of humans was created ex nihilo, by an evolutionary process, or by some other means is fundamentally not relevant to the Mormon position. Thus, although scientifically germane, theologically the entire creation process, whether literal or developmental, becomes basically a nonissue for Mormons.

## Neoliteralist Theology and Modern Science

Virtually all examples of tension between proponents of neoliteralism and those advocating a positive scientism within Mormonism deal with questions of the creation, the age of the earth, evolution, and biblical interpretation. The earliest major confrontation between these groups culminated at Brigham Young University in 1911, when several faculty members who had attempted to use the latest in secular learning, including evolutionary ideas and biblical higher criticism, to address questions of religious and theological interpretation were dismissed.[24] In 1908 BYU president George H. Brimhall had hired two pairs of brothers—Joseph and Henry Peterson and Ralph and William Chamberlin—to assist in establishing academic respectability at the university. A confrontation gradually emerged between these newly appointed faculty and several claiming to be the guardians of Mormon orthodoxy, principally Horace Cummings, church superintendent of education, and eventually Brimhall himself. Cummings was an uncompromising opponent of biological evolution and of modernist religious ideas generally, and he perceived the use of these intellectual tools as an attack on the faith of the youth of the church. Even though 80 percent of the students at BYU both privately and publicly supported the threatened professors, the Petersons and Ralph Chamberlin were dismissed, ostensibly not for teaching "heretical" views but for disobedience to their ecclesiastical and educational authorities.

In the April 1911 issue of the *Improvement Era* President Joseph F. Smith addressed this crisis in church higher education, noting that the three professors were "eminent scholars, able instructors, men of excellent character, . . . nevertheless, as teachers in a Church school they could not be given the opportunity to inculcate theories that were out of harmony with the recognized doctrines of the Church."[25] Although the issue of evolution was explicitly ignored in the *Era*, President Smith commented in an editorial in the *Juvenile Instructor* for the same month that the church's position on evolution was neutral: "In reaching the conclusions that evolution would best be left out of discussions in our church schools, we are deciding a question of propriety and not undertaking to say how much of evolution is true or how much false."[26]

Although the 1911 BYU crisis was significant, it did not primarily deal with ecclesiastical church authorities. Notwithstanding the Brigham Young–Orson Pratt controversies, the most protracted and bitter battle among Mormon church authorities may have occurred in the late 1920s and early 1930s surrounding B. H. Roberts's manuscript "The Truth, the Way, the Life," completed in 1928.[27] Although ultimately Roberts's manuscript was an attempt to construct a coherent view of God and humankind within the whole cosmic context of human existence, Roberts's exploration of gospel principles and the best of modern secular learning was in a long tradition, dating to Orson Pratt in the 1850s, of discussing science and Mormonism. As we have seen, Roberts's chapters on cosmology and astronomy developed the plurality-of-worlds idea within Mormonism's larger vision of human experience.

After Roberts submitted his manuscript for church consideration, the reading committee of the Council of the Twelve and, later, the council itself concluded that, were Roberts to press for publication, his manuscript would need some revision. Most chapters, including those dealing with a very old earth, were well received or at least not particularly controversial among church authorities. Some, however, found the chapters dealing with pre-Adamites totally unacceptable.[28]

To Roberts the evidence, from both science and scripture, for the existence of plants, animals, and various human groups was overwhelming. The problem, therefore, was to reconcile the paleontological and geological evidence while addressing the scriptural interpretation that Adam was the first man in the earth and that only following his fall did death enter into the world. Roberts affirmed that Adam was the first man of the *human* race, but he also believed that there were nonhumans who had lived upon the earth prior to the Adamic Dispensation.[29] The "lone and dreary" world spoken of in Genesis was scriptural evidence to him that the earth had been cleansed of all pre-Adamic life by some cataclysmic event that left only the paleontological remains found by modern science. It thus needed replenishing from the Adamic stock. Roberts also adopted an ambiguous view of evolution, but it was principally his pre-Adamic theory that some authorities found objectionable.

Among the few authorities who disagreed violently with Roberts was Joseph Fielding Smith, a member of the original committee appointed to read his manuscript. Smith, who even went so far as to attack Roberts's position publicly in an address to the April 1930 Genealogical Conference, denied the pre-Adamite theory and the view that there was death in the world before the fall of Adam.[30] Eventually James E. Talmage entered the debate between Roberts and Smith, arguing in favor of an old earth, cautiously accepting the possibility of pre-Adamites, but denying speciation and evolution in general. When Roberts and Talmage both died in 1933, the controversy over Roberts's still unpublished manuscript waned.

Although debate on these topics has not since resurfaced at the general church level, the opposing views of Roberts and Smith were eventually reaffirmed respectively by John A. Widtsoe in the late 1930s and 1940s and by Smith again in the 1950s. In a series of widely read articles in the *Improvement Era*, Widtsoe argued for the old-earth view and that the "days" spoken of in Genesis were periods of indefinite length. On the subject of evolution, Widtsoe affirmed the First Presidency's opinion that evolution by descent is one of several possible interpretations, thus leaving the issue both respectable and open.[31]

During the 1950s apostle Joseph Fielding Smith led a reactionary campaign to excise, once and for all, any scientific—particularly evolutionary—thinking within Mormonism. With the support of other literalist general authorities, in 1954 Smith published his influential and militantly anti-evolution, antiscience book *Man: His Origin and Des-*

*tiny*.[32] Smith recognized in Widtsoe—as well as in Joseph F. Merrill, who were both scientists and junior to Smith in the Twelve—a formidable "adversary" whom he did not wish to engage directly. The material Smith collected to publish in *Man* was fully available to him decades earlier; however, Smith, who was younger than Widtsoe and who utterly lacked all scientific training, most likely waited to publish his book until after Widtsoe's death (in 1952; Merrill also died the same year).

Not only did Smith reject old-earth geology, pre-Adamic paleontology, and evolution, but he argued that a literal reading of the scriptures (mostly the Old Testament) was necessary (and sufficient) to understand all questions dealing with origins, including the age of the earth, the antiquity of humans, and the origin of life generally. In the first chapters of *Man*, Smith outlined his overall strategy: true science and true religion will never conflict. The nature of true science, of course, is first to be understood and guided scripturally: thus, taking his lead from the Inquisition of the Galileo controversy three-and-a-half centuries earlier, Smith argued that the scriptures tell us not only how to go to heaven but how the heavens go.[33]

David O. McKay, president of the church at the time that Smith published his book, repeatedly disavowed *Man: His Origin and Destiny*—in personal correspondence but only ambiguously in public—as either an authoritative statement or church doctrine. "On the subject of organic evolution," wrote President McKay, "the Church has officially taken no position. The book *Man: His Origin and Destiny* was not published by the Church, and is not approved by the Church."[34] Smith's views on evolution, the creation (geology), and science and religion, however, have all been given wide press in the unofficial, yet widely read, compendium *Mormon Doctrine*, published in 1958 by Smith's son-in-law Bruce R. McConkie, one of the most influential neoliteralist theologians.[35] Shortly after McConkie's book was published and because McConkie wrote his book "in an authoritative tone of style" with neither church support nor direction from the First Presidency, McKay asked apostles Mark E. Petersen and Marion G. Romney to review *Mormon Doctrine*. In their reports Petersen noted over 1,000 "doctrinal errors" and Romney cited nearly forty "problem areas." The First Presidency concluded that the book is "full of errors and misstatements, and [that] it is most unfortunate that it has received such wide circulation."[36] As with Smith's *Man*, President McKay also disavowed *Mormon Doctrine* as either authoritative statement or church doctrine, indicating that a second edition would never be published, but despite McKay's injunction, in 1966 McConkie published an updated edition of *Mormon Doctrine*.[37]

## Mormonism and 'Scientific Creationism'

Although the vocal rhetoric has died down considerably since the publication of Smith's book, now that *Mormon Doctrine* has become for many Mormons (un)official *Mormon* doctrine, this neoliteralist theology based partially on a reactionary science continues to persist. These reactionary elements within Mormonism parallel a similar reaction among ideologically conservative evangelical Christians. In the Mormon tradition, these views were given their current shape by Joseph Fielding Smith's *Man: His Origin and Destiny*. In turn Smith heavily relied on the thinking of anti-evolutionists, including George McCready Price, the Seventh-day Adventist whose ideas formed what has become known today as scientific creationism.[38]

Within twenty years of the publication of Charles Darwin's *Origins of Species* in 1859, there remained in North America only two prominent naturalists, John W. Dawson of McGill University and Arnold Guyot of Princeton, who had not embraced Darwin's theories but instead remained biblical 'special creationists.'[39] The proportion of scientists, both believers and skeptics, adopting evolutionary views now as then has remained roughly the same. In the late nineteenth century, however, conservative evangelical Christians had promoted a biblical literalism and belief in 'special creation'—to wit, the view that God created all species, as well as the earth, the solar system, the stars, and the entire universe in a "special" act of creation of six days (occasionally lengthened to 6,000 years) duration.

Among evangelical Christians who rallied around this view was William Jennings Bryan, thrice Democratic candidate for the presidency of the United States and lay Presbyterian preacher. Bryan's creationism became nationally publicized during the 1925 "monkey trial" of John Thomas Scopes, a high school teacher in Dayton, Tennessee, who had violated the state's recently passed law banning the teaching of human evolution.[40] The most long-term influential creationist among evangelical Christians was not Bryan, however, but the "geologist" George McCready Price. Price based his religious scientism on the views of the Seventh-day Adventist prophetess Ellen G. White. White claimed that she had been present when God created the World, and therefore she could attest that the Noachian flood accounted for the fossil record on which geologists and evolutionists based their theories. Price published many books supporting the creationist dogma, including his influential *New Geology* (1923) as well as *The Phantom of Organic Evolution* (1924) and *The Geological-Ages Hoax* (1931), on all of which Smith based much of his thinking in *Man: His Origin and Destiny*. Following a lull in these activities during the 1930s, '40s, and '50s, creationism was revived in 1961 with the publication of Henry M. Morris and John C. Whitcomb, Jr.'s *Genesis Flood*, the most impressive contribution to strict creationism since the publication of Price's *New Geology*.

Primarily under the influence of Price and later Morris and Whitcomb, creationism, which hitherto had been almost exclusively a religious dogma, was fashioned into a pseudoscientific view. In 1963 Morris and others started the Creation Research Society, and in 1972 Morris founded the Institute for Creation Research, currently the most influential creationist organization in America. In order to counter what they perceived as the evolution threat, Morris, his associate Duane T. Gish, and the ICR changed the tactics of creationists from a biblical assault on secular thinking to a "scientific" assault on geological and evolutionary ideas. Because the American public lost interest in and failed to embrace strict creationist arguments after Darwin, it was Morris's idea that giving creationism a scientific veneer—by calling it "scientific"—would, ironically, give creationists a legitimate platform from which to combat evolutionary views.

Although masquerading as science, scientific creationism is neither good science nor good religion.[41] It is the effort by some ideologically conservative evangelical Christians to promote a view of the Bible that totally rejects virtually all biblical understanding within both normative Christianity and mainstream Mormonism. For those who have gone beyond strict creationism and embraced its offspring, scientific creationism demands that its adherents adopt a profoundly antiscience posture. Because Joseph Fielding Smith's *Man* is laced heavily with references to special creationists such as Price, it comes as no surprise that his arguments parallel those found today within the scientific creationist movement.

When neoliteralist Mormons adopt these ideas, they must realize they place themselves squarely into the evangelical Christian camp; however, they nearly always deny, overlook, or forget these intellectual roots. Creationist writings in Mormonism have persisted well after the publication of Smith's *Man*. For example, the current Old Testament manual published by the Church Educational System (CES), presents an examination of "young-earth" geology—the view that the earth was formed in 6,000 years rather than 4.5 billion years. Beginning by citing various scriptures as supporting evidence—Abraham 3:2–4, Psalm 90:4,

2 Peter 3:8, Abraham Facsimile no. 2 (figures 1 and 2), and D&C 77:12—
the manual then adds some commentary:

> Although the majority of geologists, astronomers, and other scientists be-
> lieve that even this long period [6,000 years] is not adequate to explain the
> physical evidence found in the earth, there are a small number of reputable
> scholars who disagree. These claim that the geologic clocks are misinter-
> preted and that tremendous catastrophes in the earth's history speeded up
> the processes that normally may take thousands of years. They cite evidence
> supporting the idea that thirteen thousand years is not an unrealistic time
> period.[42]

This statement is loaded with a large number of highly problematic
terms and phrases. For instance, there are no "reputable scholars" in
the field who endorse this view; an understanding of the kinds of
"tremendous catastrophes" needed to support this view is utterly lack-
ing; and, of course, a "long period" of 6,000 (or even 13,000) years is
insignificant and, by contemporary, geological standards, completely
unrealistic. Whereas the six-day view is dispensed (correctly!) in short
order in the CES manual, the 4.5 billion-year theory, which represents
solid mainstream science, is equally dismissed as being virtually antireligious—and therefore wrong.[43] The manual then continues with a
most extraordinary endorsement: "Immanuel Velikovsky ... wrote
three books amassing evidence that worldwide catastrophic upheavals
have occurred in recent history, and he argued against uniformitarianism, the idea that the natural processes in evidence now have always
prevailed at the same approximate rate of uniformity."[44]

During the 1950s and '60s, the non-Mormon Velikovsky, trained as
a physician, was one of the most extreme, anti-establishment "biblical
scientists" in America. Velikovsky assumed a literal reading of Genesis
and postulated the existence of a *war* among the planets of the solar
system in order to provide a "scientific" account of the Noachian flood
and other Old Testament events. To give added support to Velikovsky's
views and the "young-earth" theory, in addition to noting Velikovsky's
three books, the CES manual also cited two relatively well-known
publications familiar to the neoliteralist Mormon audience. None of
these sources, including the three books by Velikovsky, however, come
from mainstream science; they all represent a reactionary view of consensus science. Although one can understand the inclusion of the latter
two because of their Mormon connection, it is quite extraordinary that
CES would cite as their main source three books by a highly inflam-
matory, reactionary, self-styled "scientist" who had already been dismissed as a charlatan by the entire scientific establishment.[45]

This neoliteralist theology persists among some influential contemporary church authorities. For example, in his five-volume series on the life and mission of Jesus Christ, Bruce R. McConkie rendered a most remarkable statement to the effect that the world (earth? universe?) is at least 2.555 billion years old.[46] This value is about half the age of the earth and about one-fifth the age of the universe as presently conceived by modern science. How did McConkie arrive at this figure? He did not calculate it himself; he found it in a source in early Mormon history. Writing to William Smith (Joseph Smith's brother), W. W. Phelps noted in 1844: "[E]ternity, agreeably to the records found in the catacombs of Egypt, has been going on in this system (not the world) almost 2,555 millions of years."[47] McConkie notes that Phelps's statement that the 'world'—which McConkie interprets to mean the 'universe,' although no justification is given for this rendition—is two-and-a-half billion years old represents an "authentic account, which can be accepted as true."[48] Presumably by "authentic" and "true" McConkie means that Phelps's statement is a revelatory insight that is binding on Mormons. But how did Phelps, and later McConkie indirectly, obtain this figure? Using neoliteralist logic, it is quite easy. (See figure 8.1.) Given the fact that this numeric value is based on *fixed*

---

(1) Each creative period is the equivalent of 1,000 years of *God's* time (not humankind's) and because the *Creation* required seven periods (including one of rest), therefore the Creation required 7,000 years of God's time:

or:  7,000 years (God)/creation (God)  (C1)

(2) The reckoning for God's time is similar to that of human's time:

or:  365 days (God)/1 year (God)  (C2)

(3) 1,000 years in human's time is equivalent to one day in God's time (Abr. 3:4; 2 Peter 3:8):

or:  1,000 years (humans)/1 day (God)  (C3)

Finally, multiplying C1 by C2 by C3, or

[7,000 years (God)/creation (God)]
× [ 365 days (God)/1 year (God)]
× [1,000 years (humans)/1 day (God)]

we get:  2,555,000,000 years (humans)/creation (God)

---

Fig. 8.1—McConkie's Estimate of the Age of the World

and unalterable values, hence the result itself can never change, one can only wonder how one would calculate the age of the world in, say, a few millions of years hence.

Although McConkie later may have wavered on his earlier advocacy of a young earth, as much as one might wish otherwise McConkie was not a closet "old-earth" geologist.[49] Indeed, in the context in which he was writing, McConkie was applying this value to the age of the present universe, which, at best, makes him an "old-universe" advocate—but not for *scientific* reasons. Rather, he was engaging in what is called scriptural numerology, a field of speculation that has a history stretching back at least 4,000 years to Egypt, Babylonia, and more recently to the Pythagorean era of pre-Socratic Greece. McConkie was a popular Mormon scriptorian; he was not known, however, for his proscience views. His kind of scriptural speculation belongs in an earlier era where scholastic numerology was effective and appreciated.

## Mormonism and Modern Science

Despite repeated efforts by several church presidents, this neoliteralist tradition continues among some Mormon authorities and laypersons.[50] Still, this trend has not remained unchallenged within the institutional church. A few years after the appearance of Smith's *Man*, Bookcraft, a church-related publisher, felt compelled to answer Smith's antiscientism with a book containing the "writings and talks by prominent Latter-day Saints scientists on the subjects of science and religion."[51] Among those represented in *Science and Your Faith in God* were John A. Widtsoe, Joseph F. Merrill, and Henry Eyring, the latter easily one of the greatest scientists in America at the time and brother-in-law to Spencer W. Kimball, later president of the Mormon church. All the authors affirmed that "the facts of science, rightly looked at and understood, are helpful to the development of a sound religious faith."[52]

More recently, in an issue of the *Ensign*, the official, primary publication of the Mormon church, under the section entitled "I Have a Question" appeared the query: "Do we know how the earth's history as indicated from fossils fits with the earth's history as the scriptures present it?"[53] The editors of the *Ensign* asked Dr. Morris Petersen, professor at Brigham Young University and Mormon stake president, to provide an answer. Petersen, a professional geologist, is an advocate of the "old-earth" view. His answer was straightforward and, in a word, flew right into the face of both the six-day and the 6,000-year interpretations of the creation of the world advocated by neoliteralists. Although Petersen's answer does not constitute Mormon doctrine, the mere fact that an article sympathetic to—or at least indirectly supportive of—evolution (the word is never mentioned in his article, however) appeared in the major house organ of the Mormon church suggests that there are some powerful forces in the institutional church who understand fully the inherent dangers of the neoliteralist thinking vis-à-vis contemporary science.

Apparently, Petersen's answer was quite reassuring to many Mormons. For example, the editors printed a letter from a Mormon family who had used Petersen's article "for a family home evening lesson, and . . . the basis of an excellent discussion with our children."[54] More importantly, this episode demonstrates, as with Talmage, Widtsoe, Roberts, and others in the early years of this century, that today many at the highest levels in the church do not feel threatened by modern science and scholarship. Except for fundamentalist, evangelical Christians and reactionary Mormons, much of Mormonism and the rest of Christendom had made peace (which is not necessarily the same as acceptance) with old-world geology and most of evolution decades ago.

How could such an article be published in the *Ensign*? Is this an indication of the re-emergence of a new and more broadly conceived agenda? Actually, to their complete credit, it was the *Ensign* editors who had initiated the contact with Petersen and requested that he write the article, which then took six months to move through various channels of staff and general authorities before being published. But this makes it even more noteworthy, because it was not just sneaked into the magazine. The editors had reportedly accumulated a huge file of questions from readers asking for help with this and related scientific questions. Finally, they felt they just had to respond. The editors changed hardly a word of the original draft; the substance of the article is identical to Petersen's original intent.

Can Mormonism live peacefully with evolutionary views, or at least with a position that leaves the issue open? Apparently Joseph F. Smith, Heber J. Grant, David O. McKay, Spencer W. Kimball most recently, and many lesser Mormon authorities have suggested that Mormons should be able to do so.[55] These and countless Mormon scientists and laypersons understand, however vaguely, that science is a process and not a set of answers. Even though historically Mormonism claims that true science will never conflict with true religion, it does not necessarily follow that evolution is also the *true* scientific description of reality. As a highly integrated, coherent set of theories, however, evolution is the chief organizing principle that describes our present understanding of much of the biological world. Remove evolutionary thinking—

whether in the biological or physical sciences—and one must return to the eighteenth century.

There is a much deeper issue present, however, than whether or not Mormonism can live peacefully with various scientific views. That issue deals fundamentally with hermeneutical principles. Not unlike the Roman Catholic church during the Galileo affair, when the church reasserted its historical prerogative on matters of theological explication, neoliteralist elements within Mormonism, particularly beginning with the Roberts-Smith affair, have felt threatened by those who would interpret (Mormon) theology by adopting as legitimate a hermeneutic that gives some credence to modern historical and scientific scholarship. Historically, Mormonism had understood its theology mostly by relying on a literal hermeneutic deeply shaped by prophetic insight. The concern which Roberts, Talmage, Widtsoe, and Merrill felt in the early years of this century and which increasing numbers during the last decades now pose is a direct challenge to an older hermeneutic whose vitality has become rigid and reactionary.

### NOTES

1. Astrophysicist Robert Jastrow has recently argued precisely this cosmological position in his book *God and the Astronomers* (New York: Warner Books, 1978). For a recent and very accessible discussion of these ideas, see Stephen W. Hawking, *A Brief History of Time: From the Big Bang to Black Holes* (New York: Bantam Books, 1988). For one Mormon response to possibilities of an ex nihilo big-bang creation, see Keith E. Norman, "Mormon Cosmology: Can It Survive the Big Bang?" *Sunstone* 10, no. 9 (1985), 19–23.

2. In an address to BYU students, apostle Russell M. Nelson has recently declared that the sciences of cosmogony, evolution, and geology may be antithetical to Mormonism: "Through the ages, some without scriptural understanding have tried to explain our existence by pretentious words such as ex nihilo (out of nothing). Others have deduced that, because of certain similarities between different forms of life, there has been a natural selection of the species, or organic evolution from one form to another. Many of these people have concluded that the Universe began as a 'big bang' that eventually resulted in the creation of our planet and life upon it. To me, such theories are unbelievable! Could an explosion in a printing shop produce a dictionary? It is unthinkable! Even if it could be argued to be within a remote realm of possibility, such a dictionary could certainly not heal its own torn pages or renew its own worn corners or reproduce its own subsequent editions!" See his "The Magnificence of Man," *Ensign* (January 1988), 64–69, esp. 67.

3. Leonard Arrington, "The Intellectual Tradition of the Latter-day Saints," *Dialogue* 4, no. 1 (Spring 1969), 13–26; and Davis Bitton, "Anti-Intellectualism in Mormon History," *Dialogue* 1, no. 3 (Autumn 1966), 111–34.

4. O. Kendall White, *Mormon Neo-Orthodoxy: A Crisis Theology* (Salt Lake City: Signature Books, 1987).

5. See Bernhard W. Anderson, "Creation," and Theodore H. Gaster, "Cosmogony," in *The Interpreter's Dictionary of the Bible,* Vol. 1, ed. George Arthur Buttrick et al. (New York: Abingdon, 1962), 725–32, 702–9.

6. For a nonpolemical example of a traditional LDS position, see F. Kent Nielsen, "The Gospel and the Scientific View: How Earth Came to Be," *Ensign* 10, no. 9 (September 1980), 67–72. For an original scientific interpretation of a sequential creation account by a well-known Mormon geologist, see William Lee Stokes, *The Creation Scriptures: A Witness for God in the Scientific Age* (Salt Lake City: Starstone, 1979).

7. Conrad Hyers, "Biblical Literalism: Constricting the Cosmic Dance," in *Is God a Creationist?* ed. Roland M. Frye (New York: Scribner's, 1983), 101. For a variation on this theme by a Mormon scholar, see Benjamin Urrutia, "The Structure of Genesis, Chapter One," *Dialogue* 8, nos. 3/4 (Autumn/Winter 1973), 142–43.

8. Keith E. Norman, "Adam's Navel," *Dialogue* 21, no. 2 (Summer 1988), 85–88, esp. 81–97.

9. Ibid., 88.

10. For a (nontraditional) Mormon discussion of this point from a truly cosmic perspective, see Hugh Nibley, "Before Adam," in *Old Testament and Related Studies: The Collected Works of Hugh Nibley,* ed. J. W. Welch, G. P. Gillum, and D. E. Norton, (Salt Lake City: Deseret Book, 1986), 1:49–85.

11. The range of LDS hermeneutics from literalism to criticism is explored in Anthony Hutchinson, "LDS Approaches to the Holy Bible," *Dialogue* 15, no. 1 (Spring 1982), 99–124. Also see the response and rejoinder by James Faulconer, "Hutchinson Challenged," *Dialogue* 16, no. 4 (Winter 1983), 4–7; and Anthony Hutchinson, "Round Two on Biblical Criticism," *Dialogue* 17, no. 2 (Summer 1984), 4–6.

12. Anderson, "Creation," 728.

13. For a discussion of this possibility, see Karl C. Sandberg, "Knowing Brother Joseph Again: The Book of Abraham and Joseph Smith as Translator," *Dialogue* 22, no. 4 (Winter 1989), 17–37.

14. For the traditional Mormon account of the creation story, see Frank Salisbury, *The Creation* (Salt Lake City: Deseret Book, 1976), 50–109; for an assessment of the LDS creation narratives that takes into account contemporary (non-Mormon) scholarship, see Anthony Hutchinson, "A Mormon Midrash? LDS Creation Narratives Reconsidered," *Dialogue* 21, no. 4 (Winter 1988), 11–74.

15. See Garland E. Allen, *Life Science in the Twentieth Century* (New York: John Wiley, 1975), 1–19; Robert C. Olby, *Origins of Mendelism* (New York: Schocken, 1966). On Darwin, Darwinism, and its social context, see David Kohn, ed., *The Darwinian Heritage* (Princeton: Princeton University Press, 1985).

16. Joe D. Burchfield, *Lord Kelvin and the Age of the Earth* (New York: Science History Publications, 1975), 72–73 and passim.

17. This creation event became highly popularized by the seventeenth-century Anglican minister Bishop Ussher; see A. M. Alioto, *A History of Western Science* (Englewood, N.J.: Prentice-Hall, 1987), 270–71.

18. See Michael T. Ghiselin, *The Triumph of the Darwinian Method* (Berkeley: University of California Press, 1972), passim; and David L. Hull, *Darwin and His Critics: The Reception of Darwin's Theory of Evolution by the Scientific Community* (Cambridge, Mass.: Harvard University Press, 1974), 3–77.

19. Cynthia E. Russett, *Darwin in America: The Intellectual Response, 1865–1912* (San Francisco: Freeman, 1976).

20. Joseph F. Smith, John R. Winder, and Anthon H. Lund (First Presidency of the Church of Jesus Christ of Latter-day Saints), "The Origin of Man," *Improvement Era* 13 (November 1909), 75–81, reprinted in *Millennial Star* (18 November 1909): 721–26.

21. Joseph F. Smith, *Improvement Era* 14 (April 1910), 570. Also see Widtsoe, *Joseph Smith as Scientist*, 103–14, and W. H. Chamberlin, *An Essay on Nature* (Provo: privately printed, 1915), passim.

22. Heber J. Grant, Anthony W. Ivins, Charles W. Nibley (First Presidency of the Church of Jesus Christ of Latter-day Saints), *Millennial Star* 87 (1 October 1925), 632–33.

23. For an essential Mormon assessment of evolution, see Duane Jeffrey [sic], "Seers, Savants, and Evolution: The Uncomfortable Interface," *Dialogue* 8, nos. 3/4 (Autumn/Winter 1973), 41–75; and idem, "Seers, Savants and Evolution: The Continuing Debate," *Dialogue* 9, no. 3 (Autumn 1974), 21–38.

24. Richard Sherlock, "Campus in Crisis—BYU, 1911," *Sunstone* (January/February 1979), 11–16; and Gary James Bergera and Ron Priddis, *Brigham Young University: A House of Faith* (Salt Lake City: Signature Books, 1985), 134–48.

25. Joseph F. Smith, "Theory and Divine Revelation," *Improvement Era* (14 April 1911), 548–51.

26. Joseph F. Smith, "Philosophy and the Church Schools," *Juvenile Instructor* (April 1911), 208–9.

27. See Gary J. Bergera, "The Orson Pratt–Brigham Young Controversies: Conflict within the Quorums, 1853–1868," *Dialogue* 13, no. 2 (Summer 1980), 7–49. For a thorough analysis of the B. H. Roberts controversy, see Richard Sherlock, " 'We Can See No Advantage to a Continuation of the Discussion': The Roberts/Smith/Talmage Affair," *Dialogue* 13, no. 3 (Fall 1980), 63–78; and Jeffrey E. Keller, "Discussion Continued: The Sequel to the Roberts/Smith/Talmage Affair," *Dialogue* 15, no. 1 (Spring 1982), 79–98.

28. B. H. Roberts, "The Truth, the Way, the Life: An Elementary Treatise on Theology" (1928), chap. 30, "The Earth-Life of Man Opened," contains the problematic material and is the one still questioned by some authorities.

29. Also see Nibley, "Before Adam," 82–83.

30. Joseph Fielding Smith, "Faith Leads to a Fullness of Truth and Righteous," *Utah Genealogical and Historical Magazine* 21 (October 1930), 145–58.

31. See John A. Widtsoe, "How Old Is the Earth," *Improvement Era* 41 (December 1938); "How Did the Earth Come into Being," ibid. 42 (February

1939); "What Is the Origin of Life on Earth," ibid. 42 (March 1939); "To What Extent Should the Doctrine of Evolution Be Accepted," ibid. 42 (July 1939); and "Were There Pre-Adamites," ibid. 51 (May 1948). All except the last article are collected in John A. Widtsoe, *Evidences and Reconciliations*, ed. G. Homer Durham (Salt Lake City: Bookcraft, 1960).

32. Joseph Fielding Smith, *Man: His Origin and Destiny* (Salt Lake City: Deseret Book, 1954). For a discussion of the profound disagreement between Henry Eyring, perhaps the greatest Mormon scientist and one of the world's leading chemists, and Smith on the issue of science and Mormonism, see Steven H. Heath, "The Reconciliation of Faith and Science: Henry Eyring's Achievement," *Dialogue* 15, no. 3 (Autumn 1982), 87–99.

33. In the events surrounding Galileo's first encounter with the Inquisition in 1616, Cardinal Cesare Baronius, representing biblical moderates disavowing a literal rendition of scripture three centuries before Smith's time, affirmed the view that "the Holy Ghost intended to teach us how to go to heaven, not how the heavens go." See Jerome Langford, *Galileo, Science and the Church* (Ann Arbor: University of Michigan Press, 1971), 65. Three hundred and fifty years later Pope John Paul II adopted this same phraseology and argument; see "Address to the Pontifical Academy of Science on 3 October 1981," in *Voices for Evolution*, ed. B. McCollister (Berkeley: National Center for Science Education, 1989), 62. I am indebted to Duane Jeffery for bringing this reference to my attention.

34. William Lee Stokes, "An Official Position," *Dialogue* 12, no. 4 (Winter 1979), 90–92; also Bergera and Priddis, *Brigham Young University*, 152–57.

35. Bruce R. McConkie, *Mormon Doctrine* (Salt Lake City: Bookcraft, 1958).

36. Bergera and Priddis, *Brigham Young University*, 159.

37. On President David O. McKay's disavowal of Bruce R. McConkie's *Mormon Doctrine* as authoritative, see Marion G. Romney to David O. McKay, 28 January 1959 (copy of letter in possession of the author). Also see David O. McKay, personal journal, 7–8 January and 27–28 January 1960, in which McKay notes that McConkie agrees never to reprint *Mormon Doctrine* in a second edition (copy in possession of the author).

38. For details described herein, see Ronald L. Numbers, "Creationism in Twentieth-Century America," *Science* 218 (5 November 1982), 538–44; and Ronald L. Numbers, "The Creationists," in *God and Nature: Historical Essays on the Encounter between Christianity and Science*, ed. David C. Lindberg and Ronald L. Numbers (Berkeley: University of California Press, 1986), 391–423.

39. E. J. Pfeiffer, in *The Comparative Reception of Darwinism*, ed. T. F. Glick (Austin: University of Texas Press, 1974), 203.

40. Raymond Ginger, *Six Days or Forever? Tennessee v. John Thomas Scopes* (New York: Oxford University Press, 1958).

41. For one of the most accessible discussions of 'creation science,' see the collection of essays by religious, theological, and scientific specialists in Frye, *Is God a Creationist?*

42. Church Educational System, *Old Testament: Genesis—2 Samuel* 2d ed. rev. (Salt Lake City: Church of Jesus Christ of Later-day Saints, 1981), 28.

43. Geographer Donald W. Patten provided a 6,000-year theory to account for the Noachian flood by positing the existence and subsequent collision of the Earth with an "ice" planet hurtling through space; see his *Biblical Flood and the Ice Epoch: A Study in Scientific History* (Seattle: Pacific Meridian, 1966). Patton has lectured on the material in this book to very conservative evangelical audiences; see his "Cataclysm from Space: 2800 B.C.," a filmstrip which is still circulated within the Church Educational System and shown to unsuspecting Mormon students.

44. Church Educational System, *Old Testament: Genesis—2 Samuel*, 28.

45. The five sources cited are Velikovsky's *Worlds in Collision* (New York: Macmillan, 1950), *Ages in Chaos* (New York: Doubleday, 1952), and *Earth in Upheaval* (New York: Doubleday, 1955); *Science and Mormonism* (Salt Lake City, 1967) by Mormon scientists Melvin A. Cook and his son M. Garfield Cook; and Paul Cracroft, "How Old Is the Earth?," *Improvement Era* (October 1964), 827–30, 852. For an extensive analysis of this issue, see my "Mormonism and the Velikovsky Connection" (forthcoming). For a thoroughly devastating critique of Velikovsky, see Donald Goldsmith, ed., *Scientists Confront Velikovsky* (New York: W. W. Norton, 1979).

46. Bruce R. McConkie, *The Mortal Messiah* (Salt Lake City: Deseret Book, 1979), book 1, 29. Although this argument was later presented in his 1981 General Conference address to the Mormon church, McConkie's reference was deleted from its publication in the Mormon church's *Ensign* magazine, which reports all conference addresses.

47. W. W. Phelps to William Smith, *Times and Seasons* 5 (1844), 758. The phrase "(not the world)" was added to the 1844 article as originally published. It is not known who added the phrase—Phelps, the editor, or someone else.

48. McConkie, *The Mortal Messiah*, 29.

49. In an article on the nature of the creation process published shortly before his death, McConkie makes it clear that the "day" spoken of in Genesis is highly ambiguous and that it can be taken to mean "an age, an eon, a division of eternity." Although most certainly not an "old-earther," toward the end of his life McConkie made it clear that he was no longer willing to support the "young-earth" theory. See Bruce R. McConkie, "Christ and the Creation," *Ensign* 12, no. 6 (June 1982), 9–15.

50. See, for example, Boyd K. Packer, "The Law and the Light," in *The Book of Mormon: Jacob through Words of Mormon, to Learn with Joy*, ed. Monte S. Nyman and Charles D. Tate, Jr. (Provo: Religious Studies Center, 1990), 1–31. Mormon presidents Joseph F. Smith, Heber J. Grant, and David O. McKay, however, all reaffirmed the view that the sciences of origins are one possible theory; they never categorically rejected them as untrue.

51. Paul A. Green, comp., *Science and Your Faith in God* (Salt Lake City: Bookcraft, 1958).

52. Specifically citing Joseph F. Merrill, "Marvels Revealed by Science," in *Science and Your Faith in God*, 102.

53. Morris Petersen, "[Fossils and Scripture]," *Ensign* (September 1987), 28.

54. Lynn W. Hancock, "[Dinosaurs]," *Ensign* (March 1988), 80.

55. See Spencer W. Kimball, "The Blessings and Responsibilities of Womanhood," *Ensign* (March 1976), 70–72.

# 9

## Extraterrestrial Intelligence and Mormon Cosmology

A basic premise of earlier chapters is that Mormonism, coupled with both canonical and revelatory support, has embraced an epistemological realism and developed a positive, though cautious, relationship with modern science. Although occasionally critical of isolated cases, Mormons felt comfortable with a positive conception of the scientific enterprise. Cosmology, the science perhaps most central to any religious enterprise and certainly to Mormonism, has been a topic of intense discussion since the earliest years of the restoration. Indeed, cosmology subsumes the most basic religious ideas dealing with the origins of the universe and humankind's place in the cosmos. For Mormonism cosmology became codified in the theological principle of 'worlds without number' and found its most intense expression in the quintessential Mormon edifice, the temple.

The historical doctrine of the plurality of worlds—the Mormon idea of worlds without number—and the contemporary belief in extraterrestrial life all constitute a worldview that partially rejects the physical cosmology of traditional science, in which all natural phenomena can in principle be reduced to the laws of physics and chemistry. This reductionist imperative, however, does not compel us to adopt a *physical* cosmology as the basic organizing principle of nature. Rather, the reductionist view yields to a teleological view that suggests that life and intelligence, not stars and galaxies, is the endpoint of cosmic evolution. This 'biophysical cosmology,' as it has become known, assumes that life is the most basic property of the universe rather than a local accident. As Newton, Maxwell, and Einstein showed the universality of physical laws of nature, so the biophysical cosmology asserts the universality of biological principles. Thus we just may inhabit a universe where planetary systems on which life naturally emerges leading to technical civilizations is common. Perhaps most importantly, as with all cosmologies the biophysical cosmology also both describes and prescribes our place in the universe.[1]

In recent years there has emerged a scientific program of potentially vast proportions that speaks directly to this view. Known as the Search for Extraterrestrial Intelligence, in contemporary parlance, SETI is a scientific strategy that assumes the biophysical cosmology.[2] While the Mormon doctrine of astronomical pluralism supports SETI, however loosely, a growing number of scientists as SETI pessimists hold views at variance with the biophysical cosmology and, by extension, to the pluralist doctrine within Mormon theology. While other strategies may emerge, a failure of SETI—that there exist no extraterrestrial civilizations—could possibly, though not definitely, result in urging Mormons to reflect on their defense of science and their support of the biophysical cosmology. A complete SETI failure, taken seriously, would be a more critical problem than issues entailed in evolution or old-world geology, because the latter are not directly wedded to the theology of Mormonism. In contrast, the plurality of worlds is not only central to the restorationist scripture but, far more significantly, is at the very core of Mormon eschatology and belief in postmortal experience. Consequently, this issue reflects not only on Mormonism's relationship to a set of broader developments in science but also directly on its theology.

Despite the fact that an unusually large number of lay Mormons operate within the scientific community, the experience of Mormons since Talmage, Widtsoe, and Merrill has convinced those with critical awareness to remain somewhat skeptical of broad scientific conclusions, such as those entailed in a possible failure of SETI. That is, even though SETI pessimists may increasingly dominate the SETI debate, optimists within the scientific community at large—and those within Mormonism—will most likely still be able to maintain their support for the existence of extraterrestrial intelligence. Among optimists, both Mormons and others, the argument revolves around the fact of the vast size of the universe. Even given the technology needed to initiate contact with extraterrestrial sources, the extreme difficulty of sampling all possible places for the existence of extraterrestrial life will—in the minds of the optimists—render all objections moot.

Mormons have already exhibited their ability to rationalize difficult theological dilemmas. For example, Mormon archaeologists have yet to locate within North or South America any pre-Columbian site that can be definitely identified as fitting the description of Book of Mormon cultures. Although important and noteworthy attempts have been undertaken and much progress has been made, rather than causing Mormons to relinquish the historicity of the Book of Mormon, the controversy has only strengthened their resolve.[3]

In contrast to preceding chapters, in which historical and philosophical approaches are used to develop the narrative, this chapter explores in some detail and in strictly *scientific* terms a contemporary problem of relevance to Mormon theology. The scientific program investigated here is of enormous scope, because it entails the intersection of a large number of distinct scientific disciplines and because it has potential implications extending far beyond its technical boundary. This exploration exposes the weaknesses and strengths of science, both as an example of a positive scientism that may have something to say about Mormonism and as an illustration of the nature of the scientific enterprise.

## The Search for Extraterrestrial Intelligence

The plurality of worlds and extraterrestrial life are ideas that were indisputably wedded to virtually all human cosmologies. Not only are the ideas intimately associated with the great mythopoetic views of the ancients but the idea of an infinite number of worlds was a central feature of the atomists of Greek antiquity, generally regarded as the first scientific cosmologists. Even though the Aristotelian cosmology of the Middle Ages rejected the idea, some of the most influential philosophers at the time speculated on this matter. The Copernican cosmology of the sixteenth century held that planets were worlds like the earth, while the cosmologies of Descartes and Newton in the seventeenth century both adopted the view that the universe was full of planetary systems. In the eighteenth century Wright, Kant, Lambert, and later Herschel all espoused pluralism as a key proposition of their scientific cosmologies. With the emergence of Darwinian evolution in the nineteenth century and the development of the galactic cosmology of Shapley and Hubble in the twentieth, the biophysical cosmology has become a part of contemporary thinking.

As we have already seen, it would be a fundamental misunderstanding of the historical record to believe that science is simply an empirical enterprise unaffected by philosophical, theological, or mystical presuppositions. For example, numerous scientists and philosophers were deeply persuaded by such notions as the plenitude principle and the analogical argument that extraterrestrial life must exist. Even though we may wish to believe that science is science only when it avoids these influences, above all science is a *human* enterprise. Thus, although modern scientific methods to investigate the existence of extraterrestrial life were not available until about 1950, an ancient belief in a biophysical approach to the universe has profoundly dictated the shape and texture of the historical development of science.

Until the middle years of the twentieth century, the most serious attempt to *prove* the claims of the biophysical cosmology occurred with the Martian 'canal' episode championed particularly by the American astronomer Percival Lowell (1855–1916), the scion of a wealthy Bostonian family. The origins of the controversy emerged innocently in the fine observational work by the Italian astronomer Giovanni Schiaparelli during the 1870s. Beginning in 1877 Schiaparelli had observed markings on the planet Mars which he subsequently called by the Italian word *canali*, which, properly translated, means "channels."[4] In the hands of the non-Italian-speaking Lowell, however, this word was rendered "canals," suggesting, in contrast to "channels," a human-like construction. As a result, Lowell, an astute public propagandist of pluralist views, succeeded in promoting the probability of intelligent life on the Martian planet. Unfortunately, not only did this controversy confuse the question of extraterrestrial life but it failed utterly to examine the range of relevant issues that both protagonists and antagonists could agree upon. In fact, the canal controversy polarized those involved in the entire question of the existence of extraterrestrial life. In 1938 the science writer Reginald Waterfield noted:

> Now the story of the 'canals' is a long and sad one, fraught with backbitings and slanders; and many would have preferred that the whole theory of them had never been invented. Yet whatever harm was done was more than outweighed by the tremendous stimulus the theory gave to the study of Mars, and indirectly of the planets in general. Whether in a positive way to champion it, or in a negative way to oppose it, it attracted many able observers who otherwise might never have taken an interest in the planets. . . . So the pistol which Schiaparelli had so unwittingly let off, though it shocked the finer feelings of many, had undoubtedly been the starting signal of that race for discovery which the planetary astronomers are still successfully pursuing.

Not all scientists have agreed with Waterfield's assessment, however. More recently astrophysicist Carl Sagan has offered an alternate view: "It became so bitter and seemed to many scientists so profitless, that it led to a general exodus from planetary to stellar astronomy, abetted in large part by the great scientific opportunities then developing in the application of physics to stellar problems. The present shortage of planetary astronomers can be largely attributed to these two factors."

Most recently, Professor Michael Crowe, a leading scholar of the history of this debate, has noted: "Many benefits came to astronomy from the canal controversy, but the price paid by the astronomical community in loss of credibility, internal discord, methodological misconceptions, and substantive errors, as well as the efforts wasted in the observation of ambiguous detail, was far too high."[5]

During the 1950s two developments emerged that changed the entire scientific scope of the SETI debate: the development of radio astronomy, allowing for direct transmission and receiving of galactic and extragalactic signals, and the launching of Sputnik in 1957, representing the first serious effort of humans to embark on a space program for the eventual colonization of space itself. While both developments were strictly scientific, neither was motivated primarily by philosophical or religious views of the kind that had dominated the plurality-of-worlds question during earlier centuries. Thus, as *scientific* tools, science and technology could be used powerfully to explore the SETI question without recourse to subjective and metaphysical arguments.

As the SETI question now stands in the scientific community, there are those who find the possibility of extraterrestrial civilization alluring and favor an aggressive program of investigation. By far the most public proponents of this position are Carl Sagan and his colleagues Frank Drake and Bruce Murray. Furthermore, many additional well-respected scientists, particularly Philip Morrison of MIT and Michael Papagiannis of Boston University, are also in the forefront of these efforts.[6] Although not as vocal, others, unconvinced by the scientific arguments, are equally opposed to the entire project as fundamentally useless and a waste of limited resources.

At least since Darwin (if not Copernicus), modern science has adopted the reductionist view and has rejected the anthropocentric interpretation of the universe that humans are somehow the crowning glory of a providential cosmos. Modern scientists conceive of the universe as utterly neutral and ultimately reducible to the laws of physics and chemistry. Thus both the teleological and the cosmological arguments have been excised from the thinking of most scientists. Today a priori arguments in favor of SETI usually revolve around either the analogical or the plenitude arguments, replacing the word *God* or *creator* with *nature* or with an equivalent term that connotes some sort of philosophical neutrality. SETI advocates, who mostly have implicitly adopted the biophysical cosmology, still argue that the laws and principles of science can provide a sufficient explanatory basis for an affirmative answer. Those who oppose the idea of extraterrestrials, and who therefore also reject the biophysical cosmology, also reject both the analogical and plenitude arguments as insufficient. Both sides, however, employ strictly scientific arguments. If scientific arguments alone are used, then what distinguishes the supporters from the detractors? In other words, why do scientists come down so adamantly on one side of the question or the other? To a large extent, whether one is for or against SETI depends on one's a priori presuppositions, not on empirical evidences and the scientific arguments. This chapter will explore, not which scientist (among many) holds certain ideas, but what, ultimately, is the vindication of the biophysical cosmology.

## The Existence of Extraterrestrial Life

In considering current scientific thinking concerning possible contact with extraterrestrial sentient and rational beings, it will be necessary temporarily to suspend religious presuppositions, including the Mormon view that extraterrestrials, as children of the same God, would most likely be totally human-like, a position accepted by neither the majority of (non-Mormon) scientists nor, if science fiction is a reasonable barometer, the non-Mormon public at large.[7] In the scientific discussion which follows, however, to reduce the number of *scientific* assumptions about the possible nature of extraterrestrial life—in contrast to much of the highly speculative imagination of science fiction writers—we will assume that life elsewhere in the galaxy must be reasonably similar, though not necessarily identical, to life on the earth. The minimum assumptions to do so include (1) a terrestrial, rocky planet, (2) carbon-based life, and (3) advanced civilizations based on science and technology. First, however, a brief survey of our understanding of the nature and development of stars, galaxies, and the universe.

The strongest *philosophical* argument for the existence of extraterrestrials is that it would be ludicrous if, in all the vastness of the entire cosmos, the earth were the only inhabited planet with sentient life. Consider the following facts. Our solar system has nine planets, perhaps fifty natural satellites (planetary moons), billions of comets (though relatively few reach the inner part of the planetary system), thousands of asteroids, and a vast number of meteoroids. All of these objects revolve about an ordinary star (the sun). The sun itself is one of a few hundred billion stars of various sizes, brightnesses, temperatures, compositions, and ages that constitute the Milky Way Galaxy— one of several billion galaxies in the known universe. Galaxies themselves are enormous collections of stars clustered together in different sizes and shapes with vastly different brightnesses. Galaxies are further grouped together in larger clusters which, in turn, are grouped into superclusters. These superclusters are grouped even further into vastly larger "bubbles" that seem to be distinct from other bubbles of superclusters. These groupings are observable (either directly or indirectly) and likely continue up to the size of the entire universe. Our Milky Way Galaxy is approximately 100,000 light years in diameter (a light year is the distance that light travels in one year—about $6 \times 10^{12}$ miles). The size of the known universe is roughly 10 billion light years' radius; its age, approximately 15 billion years.

When these facts are combined, we find that the number of stars in the universe is roughly $10^{20}$—that is, 1 followed by 20 zeros. Now, argue the SETI proponents, of this vast number of stars, it would be unbelievable if only one planet—the earth—is inhabited by intelligent life! The issue, therefore, is not whether the universe is morally neutral and nonteleological—that is, whether or not there is a supreme being. For many scientists today, the issue is one of sheer numbers. Because $10^{20}$ is so absurdly large, to argue that humans are alone seems more a reflection of an arrogant and secular age than the possible state of the universe. But this is a relatively vague, qualitative argument based on an astronomical numerology; let us examine some scientific, quantitative factors relevant to the issue.

Stars are formed from nebulous material, mostly hydrogen gas. Given a concentration of this primordial material, whenever a passing shock front produced mostly from a nearby supernova outburst emerges, the gas begins to contract under the influence of gravitational attraction, thus becoming increasingly dense. As the density increases, the central temperature increases. If this core temperature increases sufficiently—to around twenty million degrees—thermonuclear reactions begin to burn the hydrogen in the star's deepest interior. The gravitational pressure from the mass of material and the pressure from the heat will balance one another, thus stabilizing what has now become a star. Depending on the amount of primordial material, stars come in a range of sizes, from very massive (around 50 solar masses) to very small (around .05 solar masses). (A solar mass is defined as the mass of our sun.) Very massive stars (called O-stars) have surface temperatures around 50,000 degrees and are much hotter than average-size stars (the sun is a relatively average star); therefore they burn at a faster rate. Low-mass stars (called M-stars) have surface temperatures around 3,000 degrees and burn at a relatively slow rate. Consequently, there are relatively few massive stars, because they burn up quickly, while there are a very large number of long-lasting lower-mass stars. Furthermore, because the density and pressures balance one another, stars remain in a stable state, known as the 'main sequence' (see figure 9.1), for millions and in many cases billions of years.

When stars evolve away from the 'main sequence', eventually becoming unstable, they enter various stages of behavior depending on their initial size. For instance, massive stars, known as giant and supergiant stars, go through a series of evolutionary stages in which they continue to transform hydrogen into helium, helium into carbon and oxygen, and so on until, eventually, during the very last stage, all the remaining chemical elements are formed, a process called nucleogenesis. At this point they explode and seed interstellar space with the heavier elements that had been formed within them. As second- and later-generation stars are born and evolve, they are composed, therefore, not only of the primordial material hydrogen but also of heavier elements.

Given these numeric and factual arguments, in 1961 astronomer Frank Drake of Cornell University attempted to estimate the possible number of extant civilizations in the universe that should be technically advanced enough to engage in interstellar communication and that would possess the interest to so communicate. His answer is based on a consideration of eight factors: the rate of star formation within the galaxy ($R.$); the fraction of stars resembling the sun that are not members of multiple star systems ($f_s$); the fraction of stars with planetary systems ($f_p$); the number of planets of each star with an environment suitable for life ($n_e$); the fraction of such planets upon which life develops ($f_l$); the fraction of inhabited planets upon which intelligent life arises ($f_i$); the fraction of such planets with an advanced technical civilization in the communicating state ($f_c$); and the lifetime of a technically advanced civilization in the sense previously discussed ($L$).[8]

Drake's final estimate of the frequency of technological civilizations in the galaxy is a numeric value derived from an algebraic expression relating these eight factors. Each factor expresses the rate or probability of one event. A probability value ranges from a numeric value of 0, meaning that that event can never occur, to a numeric value of 1, which connotes that that event will in fact happen (100 percent likelihood). Algebraically, Drake's equation is given by:

$$N = R. \times f_s \times f_p \times n_e \times f_l \times f_i \times f_c \times L.$$

The final value N, as the product of these rates and probabilities, expresses how many civilizations in the galaxy at any time are attempting interstellar communication. This formulation has the advantage of interrelating all the relevant factors needed for a careful examination of the *possibility* of extraterrestrial intelligent life. Drake's equation does

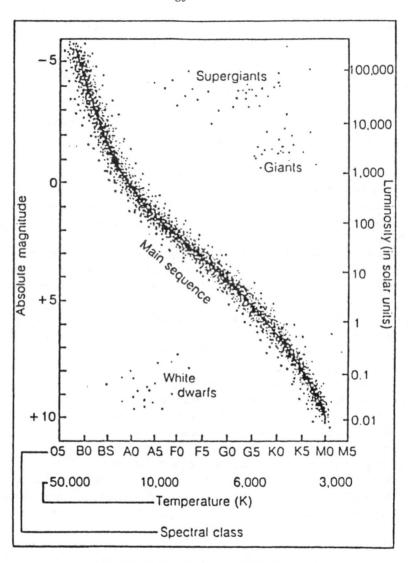

Fig. 9.1—Stellar Evolution (H–R diagram)

The 'main sequence' represents the time a star spends in the stable state after stellar birth and prior to stellar instability and death. Stars are grouped into seven general spectral classes—O, B, A, F, G, K, M—by decreasing size and temperature. The left vertical axis ('absolute magnitude') represents the intrinsic brightness of a star; the right vertical axis ('luminosity') represents the brightnesses of stars relative to that of the sun. Thus, the sun is a G2 star with surface temperature of about 5,700 degrees Kelvin and absolute magnitude of about 4.7.

*not* tell us how many advanced civilizations of the type needed actually exist. It only yields the number of civilizations that might exist within certain theoretical limits.

### Formation—R.

What is the rate at which stars are formed? Most stars in the galaxy have a mass equal to or less than the sun, and most of these have lifetimes (on the main sequence) comparable to the age of the galaxy. Therefore, because the number of known stars in the galaxy is about $10^{11}$ and the age of the galaxy is about $10^{10}$, the average rate of star formation is about ten stars per year. This number must be tempered by several considerations, however.[9]

First, some evidence suggests that during early galactic history stars formed at a greater rate than presently. Because they were composed mostly of only hydrogen and helium (the only materials available), they would therefore have been extremely poor in the heavy elements. Consequently, because none of the heavier elements needed for a solid planetary body existed at the time of their formation, the formation of terrestrial planetary systems about these stars was unlikely. Therefore, these very early stars likely do not possess solid, rocky planets and could not support life as we know it. During the continued process of stellar nucleogenesis, however, later generations of stars and planets would have increasingly been endowed with the heavy elements carbon, nitrogen, and oxygen needed for organic life-forms.

Second, the most massive stars evolve too quickly and experience stellar death before life has a chance to develop. Alternatively, for low-mass stars to provide sufficient heat and light for their rotating planetary systems, the planets must be relatively close to the star. Unfortunately, if the planet is too close to its star, the planet will experience tidal effects that will slow down, and very likely stop, the planet's rotation (as with Mercury and particularly the Moon, which always shows the same lunar surface toward the Earth). The planet's surface will then experience massive heating and cooling, precluding the emergence of life.

Finally, theoretical calculations suggest that perhaps 10 percent of all stars are in the range between high and low mass to support stable planetary systems. It has also been suggested that a large number of low-mass stars may exist, not represented on the main sequence but able to support life.

### Solar-type Stars—$f_s$

What fraction of these stars resemble the sun and are not members of binary or multiple star systems? Star formation results from the gravitational collapse (initiated by a massive shock front) of massive clouds of gas and dust called solar nebulae. Depending largely on the original amount of material, as condensation occurs a single-, binary-, or multiple-star system forms. About half the stars we observe are binary or multiple star systems whose companions are not planets but other stars.

Of the remaining single-star systems, those with giant, high-mass stars (of classes O, B, A, and F) are much hotter than the sun, and therefore they have relatively short lifetimes. Additionally, these stars probably dissipate very little of their solar nebulae, thereby forming large numbers of massive planets (of the 'Jovian' type). Stars cooler than the sun (K and M stars) dissipate their solar nebulae and therefore do not have the proper conditions for eventually sustaining life. Theoretical models confirm this view that those stars with 20 percent more mass than the sun burn too rapidly, whereas planets about those stars with 20 percent less mass freeze.

### Planetary Systems—$f_p$

What fraction of these G-type stars have planetary systems? Half the stars we observe form binary- or multiple-star systems. The likelihood of *all* binary- and multiple-star systems possessing planets in stable orbits that remain almost precisely the same distances from the suns, in order to guarantee a temperate zone, however, is theoretically unlikely. Therefore, we can reject these types of stars as possible candidates. Still, if binary and multiple star systems are common, then single stellar systems should also form (planetary) systems (nonluminous stellar objects). Unfortunately, only about 10 to 20 percent of G-type stars are single. Although astronomers have never convincingly directly observed planets about any of these systems—except, of course, for our solar system—using indirect means they surmise that many stars may possess nonluminous companions.[10] In the final analysis, however, the solar system itself offers perhaps the most compelling evidence that objects orbiting other objects, such as the gas giants Jupiter, Saturn, Uranus, and Neptune, all have large numbers of terrestrial, rocky satellites.

### Ecosphere—$n_e$

How many of these G-type stars possess planetary systems suitable for life? Although it is not entirely convincing that planets of double- and multiple-star systems would, over astronomical time, have orbits so utterly unstable as to preclude the possibility of life, we will still not include these systems.[11] Results of the Apollo, Viking, and

Voyager space missions indicate overwhelmingly that in the solar system higher life exists only on the earth. Although Mars and perhaps the Saturian moon Titan could possibly support some form of life, none has been found. In order to support life, planets must be in stable orbits that provide for a 'continuously habitable zone' (or CHZ). The CHZ is defined by its orbit and the size of a planet resident in the orbit.

Theoretical calculations suggest that a planet in an orbit that is 5 percent closer to the sun than the earth's would experience massive surface heating. This would result from the greenhouse effect, which traps the heat from the star, thus raising the surface temperature so high that life is precluded. If the orbit were 5 percent further removed from that of the earth, the planet would experience massive glaciation, because the temperatures would drop well below freezing. The history of the earth indicates that it has repeatedly experienced periods of global ice covering that, fortunately for us, have not remained permanent. Similar calculations on a planet's radius suggest that if it is smaller than the earth's by a factor of 5 percent, then not enough atmosphere will be produced to block the ultraviolet radiation from its sun. And a radius only 10 percent larger than the earth's traps too much of the outgased carbon dioxide (principally from volcanoes), thus leading once more to the greenhouse effect. Finally, even if stars are much smaller than the sun (so-called M-type dwarfs), while their luminosity would probably be too small, for Jovian type planets the greenhouse effect could produce an atmosphere with reasonable temperatures; but then these planets would not be terrestrial and the subsequent emergence of life as we know it is highly conjectural at best. In general, because giant stars evolve too quickly for life to emerge, the possible existence of ecospheres is irrelevant; while dwarf stars may have ecospheres of the type needed, many of these stars, being mostly first generation, do not possess the materials for terrestrial planets.

In the solar system there are two planets—Venus and Earth—that fit some or all of these criteria. Now let us suppose that within the terrestrial region of the solar system (roughly between the sun and Mars), we randomly place these planets. The probability that one of the planets will randomly end up in the sun's CHZ is 5 percent.

### Emergence of Life—$f_1$

How many of these life-sustaining planets will actually develop life? Again, except for the earth, we know nothing directly of other planets. This factor, therefore, is among the most hotly debated; but for purposes of discussion we will initially adopt a conservative value. There are at least five macrosteps needed for an examination of this question: (1) the formation of basic organic molecules (monomers) from very fundamental chemical elements, specifically hydrogen, nitrogen, carbon, and oxygen; (2) the development of macromolecules (polymers), specifically complex amino acids from which one can build proteins; (3) self-replicating chemical systems; (4) the emergence of unicellular organisms defined by growth and reproduction; and (5) the evolution of increasingly complex organisms in an environment defined by evolutionary mechanisms, such as natural selection.

Some scientists, particularly nonbiologists, have argued that, because of the physics and chemistry of primitive planetary environments, self-replicating molecular systems are bound to occur. And if the environmental medium is saturated with such self-replicating systems, natural selection and biological evolution will dominate the process. Because we may comfortably assume geological time-scales, the evolution of complex organisms is inevitable, they argue. Furthermore, empirical evidence from our own earthly environment strongly supports the view that, once it has arisen, life is generally very tough and resilient.

Other scientists, however—particularly biologists and chemists—contend that the mechanisms needed for the natural emergence of macromolecules and for self-replicating chemical and biological systems are either so poorly understood or so unique that it is not at all inevitable that such systems will develop. Experiments have verified that monomers can be developed quite easily under "primitive earth" conditions. Processes needed for the development of polymers and for the emergence of self-replicating chemical and biological systems are least understood. Natural selection and evolutionary mechanisms, conversely, are perhaps the most widely understood and verifiable processes, at least at certain levels of organization. Until the processes directing the development of polymers and self-replicating chemical and biological mechanisms become clearly understood, this factor is presently one of the most speculative.

### Emergence of Intelligence—$f_i$

Once life has arisen, how many of these inhabited planets will develop intelligent life? Again, we can offer only a reasonable estimate. The development of intelligence and manipulative ability itself can be expressed in a kind of evolutionary equation in which the parameters represent a large number of individual events. Although it is possible to consider many different forms of life that might hypothetically emerge and develop into intelligent species, for purposes of this discussion we will not speculate about exotic forms. Therefore we will limit ourselves to an examination of the conditions needed for the emergence of humanoid forms of intelligence. Let us assume, therefore, the following conditions.

(1) An oxygen-rich atmosphere needs to develop, because otherwise no ozone (heavy oxygen) layer would emerge to protect terrestrial life-forms from deadly ultraviolet radiation from the parent star. Furthermore, it is difficult (i.e., highly speculative) to imagine an advanced technical civilization developing without the use of fire; and fire requires oxygen (not hydrogen). (2) All higher forms of terrestrial intelligence are land dwellers. Even the dolphins and whales, whose ancestry evolved onto the land and then returned to the sea, are mammals. Moreover, environmental conditions on land are significantly more hostile than those in the water, a factor that is required for the adaptive pressures of evolution to work most effectively. (3) Cold-blooded animals are limited in their protective abilities, because they do not function well during the nighttime. Warm-blooded animals (including all mammals), however, remain alert and defensive during their entire life cycles. Natural selection being what it is, warm-blooded animals should eventually dominate their cold-blooded cousins. Therefore, the regulation of body temperature should develop. (4) The development of two eyes and two hands with opposable thumbs should provide both protection and dexterity. The overwhelming majority of species on this planet is two-eyed, suggesting that one eye is wholly inadequate while more than two is superfluous. It is also hard to imagine sentient land dwellers without an effective means of grasping and building tools. (5) Besides the discovery of fire, the development of big brains (i.e., high ratios of brain mass to body mass) and the freeing of hands are needed for the invention of a significant technology. Finally, (6) the development of complex social structures, including centralization and specialization, is needed for the emergence of stable technological civilizations.

Although intelligence, or at least technological civilization, has arisen on the earth, it is not at all certain whether intelligence as we know it would evolve again in an identical manner. The adaptive value of intelligence and manipulative capacities is so great, however, that, once started, evolution would most likely bring it about. This depends, of course, on getting favorable mutations, and that is completely independent of their adaptive value. Intelligent civilization on our planet has arisen roughly halfway during the sun's lifetime (on the main sequence). The vast majority of stars in the galaxy are older than the sun, and because intelligent civilization has only recently emerged on the earth, it is possible that, of those planets supporting life, a considerable fraction would produce intelligence and eventually a civilization.

### Advanced Civilization—$f_c$

Of such civilizations, how many will evolve into an advanced technical civilization capable of interstellar communication? Whereas terrestrial civilization achieved very modest technical abilities about 10,000 years ago, modern science and technology have developed only in the last 300 years. And this occurred only within one of the roughly twenty civilizations—namely, western European society—that have developed on the earth. Only within the last century has technology itself been largely driven by science rather than an independent crafts tradition. The development of a technical civilization seems inevitable, but whether it survives its own creation remains yet to be seen. There is no really persuasive argument that, among planets inhabited by intelligent beings, technical civilization will not eventually emerge. As a question of historical and sociological fact, once intelligence itself emerges, it seems likely that extended cultural development will result in civilizations that have invented advanced technologies, such as radio and computers.

### Civilizational Lifetime—$L$

Finally, what is the lifetime of a civilization with advanced technology possessing both the interest in and the ability to engage in interstellar communication? Human civilization has achieved this level of technical ability only within the last 100 years. Unfortunately, we have also developed thermonuclear and biochemical weapons sufficient for our complete destruction. It is certainly possible, particularly in light of the geopolitical events since 1989, that civilizations will permanently settle international political, economic, and cultural differences before weapons of mass destruction are used. Still, if other forms of civilizational destruction do not manifest, such as some exotic disease, the (unintentional) release of some deadly recombinant biological agent, or global pollution of the air or water, it is quite possible that civilizations may survive virtually indefinitely. Therefore, it seems reasonable that either (1) a technical civilization will self-destruct after roughly $10^2$ years, or (2) a technical civilization solves potentially global destructive forces soon after reaching the communicative phase. If it achieves the latter, there is no compelling reason that it will not remain socially and environmentally stable for very long periods of time, perhaps on the order of the stellar evolutionary timescale ($10^{11}$).

### Extraterrestrial Civilizations

There is a tremendous range of opinion among astronomers, biologists, and chemists, however, concerning the individual values in the Drake equation. While the value for $R_*$ and perhaps that of $f_s$ are reasonably well understood, the estimate for L is anything but precise. The remaining parameters are probability values whose likelihood of occurrence is not at all definite, well defined, or scientifically precise. Thus at least six of the eight parameters in the Drake equation are highly debatable. In order to provide the range of opinions, we will estimate both the most optimistic and the most pessimistic values:[12]

| Drake factor | Optimistic values | Pessimistic values |
|---|---|---|
| $R_*$ | 1 | 1 |
| $f_s$ | 1/10 | 1/10 |
| $f_p$ | 1/10 | 1/100 |
| $n_e$ | 1/20 | 1/100 |
| $f_l$ | 1/100 | $1/10^{10}$ |
| $f_i$ | 1/2 | $1/10^{10}$ |
| $f_c$ | 1/2 | 1/100 |
| L | $10^{11}$ | 100 |

By adopting values from various SETI specialists and critics and substituting these values into the Drake equation, we obtain as the steady-state number of existing advanced technical civilizations in the galaxy a value for N that ranges from an optimistic $10^6$ to a pessimistic $10^{-25}$.

On the one hand, this suggests that of the 100 billion stars in the galaxy, there might be approximately one million planets with advanced civilizations. The *average* distance between such civilizations, therefore, would be about 300 light years. Although this value of N indicates that only about 0.001 percent of stars would have inhabited planetary systems with advanced civilizations, the number is much greater than 1. As astrophysicist and Mormon scientist Hollis Johnson has expressed it, "Any answer different from N- would be exciting! [Indeed], it seems very likely that there are many civilizations in the galaxy around us."[13] On the other hand, a value as low as the pessimistic value reinforces what many scientists presently believe: at least one of the parameters, if not several, must be virtually zero, in which case N itself must be zero. In this case, except for Homo sapiens, there would be no extraterrestrials and no communicating civilizations outside the solar system.[14] In other words, we would be alone! And, except for humans, the universe would be utterly silent!

### Contact with Extraterrestrial Life

There are two *scientific* strategies we may use if contact with another technical civilization is to become possible: (1) actual physical contact through space travel and (2) radio communication.[15]

Space scientists and astronautical engineers believe that the fundamental concern with interstellar space travel is the feasibility of extended travel with *limited* resources, which poses the following problems. In order to decrease travel time between star systems, one must increase the speed of the spaceship close to the velocity of light. Here the principles of special relativity, which have been shown to be empirically reliable, would be utilized to shorten the trip as much as possible. As a result, the onboard occupants would experience "time dilation," in which time passes more slowly for the spaceship inhabitants, traveling at a speed near the velocity of light, than for those who stay behind. For example, let us suppose that a spaceship is sent to a star system about twelve light years distant and that during travel it reaches 99 percent the velocity of light. The crew members would age only about ten years, though the round trip would require approximately twenty-eight earth years. Although the space occupants would experience relatively little biological aging, travelers to stars much further distant (and virtually all stars in the galaxy are at vastly greater distances) would return to the earth hundreds, thousands, and even millions of years after their original departure.

This is a relatively minor inconvenience compared to the problems with actually building an interstellar spaceship.[16] Assuming that our society eventually becomes capable of solving the technical and engineering problems required for extended space travel, conservative estimates suggest that, if speeds close to the velocity of light are desired even for relatively short distances of twelve light years, a ten-ton payload (about the size of a bus) would require about 400,000 tons of fuel (in the form of matter and antimatter). Now if the spaceship must complete its travel in years, rather than centuries or millennia, then to achieve the needed accelerations the amount of radiation generated by the propulsive system is on the order of $10^{18}$ watts, equivalent to the total power the earth receives globally from the sun. But this is not sunshine: it's pure gamma radiation, the most destructive radiation in large doses. So the problem becomes one of shielding the spaceship from excessive irradiation. Additionally, even though "empty" space contains only one hydrogen atom per cubic centimeter, at speeds approximating 99 percent the velocity of light, the spaceship would encounter hydrogen atoms that appear to the space occupants like six-billion-volt protons. That's enough energy to disintegrate most terrestrial materials.[17] If speeds much slower are acceptable, say 10 percent or less of the velocity of light, then ratios of fuel to payload, as well as the radiation factors, are proportionately reduced. Of course, at slower speeds the time for interstellar travel is then considerably lengthened.

On the time scale of geological and stellar events (measured in millions and hundreds of millions of years), there is sufficient time to resolve entirely the technical problems of energetics. Therefore, if humans are willing to accept that hundreds and even thousands of earthbound generations will pass between round-trip contacts among civilizations, then interaction is within our reach. The major advantage to such interstellar (physical) connections is that once contact has been achieved (after decades, centuries, or even millennia) one does not need to wait vast periods of time for information to be exchanged directly, face-to-face so to speak, as would be the case using radio communication. Thus physical interstellar contact has two major advantages: (1) actual contact between an advanced civilization or even an intelligent, pretechnical society, and (2) the exchange of artifacts among the communities.[18] It also has several significant disadvantages: (1) because the species would have developed in radically different biological environments, it is reasonable to assume that the different species would each possess radically different antigens, thus physical contact would likely result in massive biological contamination; and (2) if human history can be trusted, the contacted society would most likely be swallowed up culturally by the superior civilization.

Aside from the culture shock of engaging in *physical* contact with an extraterrestrial civilization, which, by definition, must be quite advanced to have traveled such great distances to earth, there is a tremendous technical disparity between space travel, on the one hand, and radio or electromagnetic communication, on the other.[19] Although space hopping around the galaxy seems a lot more romantic, its infeasibility is perhaps the strongest argument favoring radio communication. Consider: a spherical volume only 300 light years in radius contains about 3,600 stars roughly similar to the sun. Therefore, the likelihood of other civilizations having already broadcast seems quite large, assuming both the existence of such civilizations and the possibility that they have been transmitting for long periods of time. Because human telegraphy is a mere century old, we are very young technically and therefore we would be more successful if we could establish contact with an advanced civilization that had already been broadcasting for much longer. Before transmitting our signals, however, we will want to listen extensively to galactic signals. Finally, because the electromagnetic spectrum is very wide, where do we start listening? In other words, where on the radio do we set the dial? Without a common frequency, two separated sources trying to locate one another would be forever invisible to one another. Although many frequencies have been proposed, as it turns out in the radio range there is exactly one universal choice. The neutral hydrogen atom, the most common substance in the universe, vibrates at a frequency of 1,420,405,750 cycles/second or a wavelength of about twenty-one centimeters. Although we might thus be reasonably convinced about the most feasible frequency of transmission, there still remain technical issues of bandwidth, direction, and related problems. For example, if scientists hope to detect a signal broadcast from a planet in a distant planetary system, the accelerated motion between the earth and this planet gives rise to a received signal that drifts in frequency.[20]

Radio communication also suffers from generational time problems not dissimilar to those involved in actual space travel. Suppose, for example, that one million advanced technical civilizations are distributed equally throughout the Milky Way Galaxy, roughly every 300 light years. Then a radio message, even to the nearest extraterrestrial civilization, and its response would require nearly 600 years. Extended conversations would obviously be much longer, and contacts with societies across the galaxy would require even longer time intervals— 10,000 to 100,000 years. In recent years, however, a number of SETI scientists with support from NASA and the Planetary Society, the largest scientific society in the world dedicated to the search for extraterrestrial intelligence, have developed techniques for multichannel scanning of millions of signals simultaneously.[21]

## Where are the Extraterrestrials?

Either way the question remains, Where are the extraterrestrials and why have they failed to contact us?[22] To date, no empirical evidence, derived from either artifacts or interstellar signals, has ever been discovered to indicate their presence. Is it therefore the case that they do not exist and possibly never have existed? A number of scientists have suggested the following possibilities to explain the lack of evidence. (1) Physical explanations: space travel is simply infeasible for reasons related to some physical, astronomical, biological, or engineering difficulty. (2) Sociological explanations: extraterrestrials have deliberately chosen not to contact us for a variety of imagined reasons. (3) Temporal explanations: there has been insufficient time for extraterrestrials to visit us using space travel. (4) Perhaps extraterrestrials have already visited us and since departed.[23]

The major obstacle to extended interstellar travel is the vast distance between stars, which requires an enormous time for the occupants of a spaceship to travel. Speeds approaching the velocity of light present serious physical difficulties for a spaceship, including the fuel-to-payload ratio and certain shielding problems. These constraints can all be resolved in a variety of ways, however. For instance, using techniques of 'suspended animation,' or freezing the spaceship occupants, or sending frozen zygotes which upon arrival would be thawed by robots to produce a living population are all theoretically possible. Because high relativistic speeds are very problematic, perhaps the most palatable solution would simply be to plan a space voyage at low relativistic speeds requiring many generations. Such voyages could quite easily engage in protracted interstellar travel, yet, over many generations— perhaps hundreds or even thousands—space travel may itself ultimately mitigate this approach. Because much of the technology is already within our reach, this latter approach requires the fewest assumptions.

The sociological explanation is far less tractable than the objections to any physical explanation for the apparent lack of extraterrestrials. What would compel extraterrestrials not to engage in interstellar travel or communication? Several possibilities have been suggested. (1) Advanced civilizations may develop consuming spiritual or aesthetic interests. (2) Not long after the discovery of atomic science, technologically advanced societies self-destruct. (3) Old civilizations eventually degenerate and lose both the interest and the means to communicate. (4) Extraterrestrials have long ago discovered the earth and its Homo sapiens population but have deliberately chosen not to contact us but merely to observe, much as in a zoo, until we mature sufficiently in social, cultural, and mental terms to merit admittance into their galactic club.[24] (5) Extraterrestrials clearly recognize the essential competitive nature of evolution and fear that they will be gobbled up by an alien civilization if detected; therefore, they elect to remain silent. (6) Contrary to our experiences, extraterrestrial civilizations develop stable, zero-growth populations that essentially mitigate the (forced) need to emigrate to, or at least explore, other planetary systems. In contrast, at the current growth rate human beings could populate one inhabitable planet about each star in the galaxy within a mere 20,000 years. At least one of these objections must apply to *every* race of extraterrestrials in the *entire* galaxy, because we have never detected, let alone encountered, even one extraterrestrial civilization. The chances of this happening on *each* inhabited planet are compounded and therefore the sociological objection must be diminishingly small.

The temporal explanation for the lack of extraterrestrials seems reasonably convincing. Suppose that at some time in the history of the galaxy one civilization emerges and eventually achieves an advanced state of technology that enables it to engage in protracted interstellar travel. Because of the rather severe generational problem, as well as the technical problems in achieving and maintaining near relativistic speeds, this civilization chooses to develop the ability to build a few machines that can perfectly self-replicate themselves under control of artificial intelligence, rather than natural intelligence. Such a machine is called a von Neumann probe after mathematician and computer pioneer John von Neumann, the first person to analyze self-reproducing machines.[25] Using even primitive rocket technology, a von Neumann probe would require about 100,000 years to cross the average distance between two stars in the galaxy. Assuming that the probe finds a planet containing all the raw materials from which it can self-replicate and allowing conservatively 1,000 years to complete the process, thus finishing with two von Neumann machines, it requires about 40 million years before the descendants from a single probe reach *every* star system in the galaxy. This is a relatively small amount of time, because a galaxy, such as the Milky Way, is at least three hundred times older. Because there are many stars whose age is in excess of six billion years (the earth's age is around 4.5 billion years), von Neumann probes should have saturated the galaxy millennia ago. Unfortunately, we do not find such machines any place on the earth. The conclusion is that they must never have been sent!

This argument is logical, not empirical, and it too is not without flaw.[26] For instance, had von Neumann probes landed and remained on the surface of the earth, then, over geologic time, the pressures due to the movement of the earth's crust (plate tectonics) may have crushed and utterly destroyed all such machines. Still, this assumption can be avoided if one of the machines remained in planetary orbit. A more devasting criticism, however, questions whether such machines can in principle be developed. While theoretically possible (it is claimed), these machines must be able to do everything (and more) that an intelligent human being is capable of doing: they must be able to reproduce themselves, to repair themselves, to launch themselves into space, and finally to travel throughout the entire galaxy for virtually an indefinite period of time. Ultimately, these machines are premised on the view that eventually artificial intelligence and very smart and capable robots will become a fact. While presently speculative, artificial intelligence is a concept that has been widely recognized as overly optimistic, overly confident, and facing insuperable practical and theoretical obstacles.[27]

At any rate, whether extraterrestrials had visited the earth during some earlier epoch is empirically highly questionable.[28] The evidence that some writers have offered suggesting that early humans or even humanoids were visited by proto-gods who biologically altered the evolutionary process is suspect at best and utterly ad hoc at worst. Of course, this argument can be settled in the affirmative as soon as some artifact is discovered that can be shown to have an extraterrestrial origin. For example, the synthesis of certain organic compounds or the development of certain kinds of materials requiring a weightless environment would be sufficient evidence of extraterrestrial visitations, because humans are yet incapable of such technological results. Others have argued that extraterrestrials have not only arrived but that they are still here. The UFO and proto-god hypotheses, however, have a prominent history of exaggeration and lack of credibility.[29]

The question of whether sending civilizations exist remains very puzzling. The temporal argument rules out only those civilizations that are capable of launching von Neumann probes and then, in fact, choose to do so. It does not rule out those civilizations that for all the sociological reasons given earlier decide not to engage in interstellar communication. If extraterrestrials have come and gone, why have they not left some kind of artifact or an unmanned probe in the solar system that would report back to them when our civilization had reached a certain level of maturity?

As far as radio signals are concerned, why have scientists not received identifiable signals from extraterrestrial civilizations? Several possibilities come to mind. (1) Signals from extraterrestrial sources exist, but we have been unable to decode the messages; in other words, we have not yet discovered the correct frequency of transmission or perhaps we have simply failed to understand some exotic, but hidden, message in the radiation transmission from the stars, or we have not looked in the right direction. (2) There has simply been insufficient time for signals to reach the solar system. And (3) no signals exist whatsoever. The first possibility is currently accepted by all SETI proponents; the second possibility suffers from sociological explanations already given; and the third possibility is favored by SETI pessimists.

Presently, no space travelers and no signals from space can be positively identified as having originated from an extraterrestrial civilization. Although we have barely begun to look, to date the ("scientific") heavens are quiet—utterly quiet!

### Life in the Universe

Since 1961, when Frank Drake proposed his equation for estimating the number of communicating civilizations, numerous scientists have entered the debate. Many aspects of this debate need to be considered to have a reasonable grasp of the complexity of the situation. Simply stating the case for SETI is well and good; but, in a word, the Drake equation is still highly conjectural. What is needed is a clear understanding of the assumptions required in formulating the equation. Whereas the early advocates of the SETI program were mostly astronomers, who by natural disposition seem almost universally in favor of SETI, others have entered the debate, particularly biologists and chemists, who have explored the nonastronomical assumptions of SETI with considerably less optimism.

A significant concern of the SETI opponents deals with the origin of life and the subsequent evolution of an intelligent civilization.[30] An immense number of factors must be required for life to emerge even in its earliest stages, and once living organisms have developed, the conditions needed to sustain them become increasingly complex. In order for stars to support planetary formation leading to the emergence of life, stars must have a particular luminosity and size. Stars in the extreme, dwarf and giant as well as cool stars, do not have the kind of ecospheres about them in which a planet must be situated in order to sustain life.[31] Therefore, ecospheres with terrestrial planets are limited mostly to the G-type stars, such as the sun. In addition, the planet itself must rotate on axes that are precisely inclined, its mass must be narrowly defined, it cannot have a highly eccentric orbit, and its moons must orbit within narrow limits of tolerance. The latter factors all seem to be required for the development of the critical weather patterns.

Moreover, we need to understand more fully certain conditions about the earth's early atmosphere and the activity of the sun, as well as the formation of complex organic molecules and macromolecules, such as proteins and nucleic acids, and how they organize into living systems in the full sense of the word. Assuming that any planet with a suitable ecosphere will eventually develop primitive organisms, much is needed for evolution to occur and life to progress and diversify. Although we know it happened at least once, the further organization of macromolecules into cellular life is still fraught with additional difficulties. In order for mutation, recombination, and selection to control the development of living organisms, these processes themselves must already have been invoked for living organisms and cellular systems to emerge from loose macromolecules. Therefore some sort of feedback and encoding must be postulated in order for the nascent organisms to diversify and adapt to their environments. Processes of adaptation, of acquiring, converting, and organizing materials and energy, are considerably different than the processes needed in synthesizing and organizing macromolecules.

Even assuming that life will emerge, or that it is the natural condition of the universe that life will appear, it is still another question whether that life will be intelligent. The evolution of life (as we see it on earth) can and does pursue so many different paths that it does not seem empirically inevitable that intelligent life with humanoid nervous systems must emerge. On earth, only one (intelligent) species capable of developing advanced technology has evolved within two large super-kingdoms of living things that include tens of thousands of evolutionary lineages. Even among animalia, with its twenty-five phyla, intelligence in the form of humans has emerged in only one species among over a hundred, in only one order among more than twenty, and in only one among over fifty classes of the chordates.

Although the earth is roughly 4,500 million years old, life on the earth began about 3,500 million years ago. Yet intelligent terrestrial life has existed only about 100,000 years—barely 0.002 percent of the earth's history. In other words, if we assume that life does evolve and if macromolecular systems have occurred elsewhere in the universe, because the fraction of intelligent life compared to all life-forms that have evolved on the earth is virtually nil, the probability of a highly intelligent species evolving elsewhere is, therefore, correspondingly extremely low. Consequently, the conditions for the origin of humanoid life seem to be extremely special.

Still, one can make an argument that, given enough time, on the order of the geologic time scale, intelligent life must evolve because intelligence can easily become one of the strongest factors in the struggle for survival among competing organisms. For example, though not necessary, big brains, hands, and tools seem to be correlated. Tools can easily be used as weapons of defense and offense. Consequently, nature would select those individuals and species that possess intelligence. Thus, as long as the environment provides these kinds of adaptive pressures, intelligence would continue to become a valuable trait. Hence, intelligence would be selected and would increase according to the principles of evolution.

Because SETI is such a complex program and because the possibilities of extraterrestrial contact are enormously appealing to SETI specialists, as well as to laypersons (witness almost universal interest in science fiction, UFOs, and other extraterrestrial topics), advocates have been very successful in furthering their work. The International Astronomical Union, the largest professional astronomical society in the world, has a division of specialists devoted to SETI. In addition to financial support from the Planetary Society, the National Aeronautics and Space Administration (NASA) is helping to fund the search for extraterrestrial signals. SETI may be a project seriously doubted by many specialists, but it has equally persuasive men and women who view it as perhaps the most exciting scientific program ever conceived.

Given some of the pessimism implied in the foregoing arguments, the tenacity of SETI advocates can perhaps best be explained more as a deep-rooted commitment to a cosmology that defines their world-view—in this case, the biophysical cosmology. Cosmologies define how people understand their relationship to nature and the cosmos; they constitute the essential core of a person's belief structure. Therefore, for many advocates, a belief in SETI has become a sort of secular religion. If contact with extraterrestrials can be established, that very fact will provide a meaning to some really basic religious yearnings about life that twentieth-century secular thinking has rejected. To wit, it will demonstrate that the emergence of life in the universe is a natural consequence of its very existence. This is a weak form of the 'anthropic cosmological principle.'

## Modern Science and the Anthropic Cosmological Principle

At least since the seventeenth century, science has suggested that life has arisen precisely because the universe is in a certain state of existence. Those inclined to deny the existence of God have argued that there simply is no need in the universe we inhabit for Providence or any kind of divine presence. In recent years this view has been turned literally upside-down by the re-emergence of the 'anthropic principle,' one interpretation of which holds that the universe is determined by the existence of human intelligence in it—i.e., because human life exists, the universe must have been determined in a very precise, ordered, and deliberate way.[32]

As a generic concept the anthropic cosmological principle has several versions. Its 'weak' form acknowledges the existence of humans in the universe (or at least life as we know it on this planet): "The physical and cosmological values of the universe are restricted by the requirements: (1) there exist places where carbon-based life can evolve, and (2) the Universe is old enough for it already to have done so."[33] Thus the 'weak anthropic principle' strongly correlates all cosmological observations to an all-encompassing 'selection effect'—namely, our own existence!

Science makes empirical observations and invents theories and laws to explain those observations (see chapter 3). At present there are many fundamental, cosmic numbers that seem to characterize particularly crucial features of the universe—for example, quantities like the electric charge of the electron, the ratio of the masses of the electron and the proton, the gravitational constant, the strength of the strong force between nucleons, and Planck's constant describing the energy increase or decrease as electrons move in their orbits. The theories and laws of science, however, do not at present allow us to predict the values of these numbers; they are discovered, sometimes serendipitously, by observation and experiment. Presently there is no 'grand unified theory' incorporating the strong and electro-weak interactions of the nucleus with gravity into a unified description of nature. Consequently, explaining the existence of everything in the universe in terms of a number of 'fundamental' parameters is far too premature to infer that the universe is fundamentally anthropocentric.

It is not sufficient that science has concluded that certain theories are "correct," however, and therefore the existence of God is assured.[34] For instance, if the electric charge on the electron had been slightly different, either stars would not be able to burn hydrogen and helium or they would not be able to explode. Or if the age of the earth were much younger, humans could not exist because evolution would not have had enough time to function. Or if the gravitational gradient had been slightly different, the universe would not have evolved with conditions favorable for the emergence of organic life. Or if the mass of the universe had been slightly greater, the big-bang theory predicts that universal gravitation would have caused the universe to collapse by now. This naturalistic view has been advocated at least since the atomists of Greek antiquity. The anthropic principle, however, states something much stronger. The very laws themselves describing the fine tuning, as it were, of the cosmic fabric were precisely prescribed *in the beginning*. Thus, following one interpretation of the 'weak anthropic principle,' it is difficult to avoid the conclusion that those laws were deliberately so established.

Ultimately, this version of the anthropic argument is tautological. Since we are here, the laws must have been thus-and-so. Since they are thus-and-so, we are here. One can argue, however, that the likelihood of *all* these laws being precisely thus-and-so simultaneously— as a matter of accident, as it were—is virtually an impossibility. In other words, the probability of these laws having *fortuitously* and *independently* emerged is essentially nil. For example, if the laws of nature (e.g., as contained in astronomy, biology, chemistry, geology, and physics) were slightly different, life as we know it could not exist, because, for one thing, the carbon atom would be too scarce. This argument, however, suffers from the *assumption* that our concepts of the world (i.e., the laws) are independent of consciousness. It is certainly possible that these laws are cognitively, and not ontologically, interdependent and that eventually they will be understood in some kind of 'grand unified theory' of the universe.

In contrast, the 'strong anthropic principle' does not necessarily assume anything about life as we know it: The universe must contain those properties that allow intelligent life to develop within it at some stage in its development.[35] The classical—or teleological—interpretation of this 'strong' version lies at the heart of virtually all historic attempts at natural theology, and as such it represents a modern version of the teleological argument writ large. Since the late Middle Ages the history of humankind has been to dethrone humans from their special place in nature. First Copernicus, then Darwin and Sigmund Freud, and now, in the last decades of the twentieth century, 'artificial intelligence' have gradually removed humans from their privileged place in a geo-anthropocentric universe. The anthropic principle would have us reverse this tradition by adopting science as the exclusive arbiter of modern times.

There are some who understand the appeal of the anthropic principle as an ill-conceived enterprise and, consequently, reject its inherent argument. As one critic of the anthropic principle has argued, the "selection effect," assumed in both the 'weak' and 'strong' forms of the anthropic principle, leads to the following: "What we observe is conditioned not only by the fact of our existence, but also by the nature and capacities of our perceptive and intellectual faculties." This is not to say that "we see the universe the way it is because we exist," but rather, conversely, "we exist because the universe is the way we observe it to be, so we could not observe it otherwise."

To speak of a selection effect is to presume a known (or, at least ascertainable) range of fact beyond what is selected. But in the case of selection exercised by our sensory and rational capacities, such talk is . . . unknowable in the very nature of the case. The presumption now is that our scientific knowledge is irremediably subjective and that the physical reality lies forever beyond our ken. Any such subjectivism presages epistemological disaster. It would imply that our science . . . was not really knowledge at all. . . . But, if our scientific observations . . . were an inevitable 'selection' resulting from the nature and limitations of our faculties, . . . they might well be no more than delusions.

Consequently, asserting a 'selective effect' must be treated with the "utmost caution and restraint."[36]

Other interpretations of the anthropic principle have also been offered, such as Wheeler's 'participatory anthropic principle' ("observers are necessary to bring the Universe into being"), a 'many-worlds' interpretation ("an ensemble of other different universes is necessary for the existence of our Universe"), and the 'final anthropic principle' ("intelligence will emerge in the Universe, and as a consequence it will never die out").[37]

Many of those discussing the anthropic principle today are not only philosophers and religionists but also scientists.[38] Indeed, most recently it has been physicists who have revived the whole conversation. In the last four centuries or so, science in the Western tradition has become almost deified, becoming for a majority of people the singular authority on matters of, for example, medicine, the environment, and the very nature of the universe itself.[39] Consequently, for many who disavow traditional religion as authoritative and institutional, science has become the religion of humanity.[40]

For others who claim some institutional religious affiliation, science, in light of the anthropic principle, shows promise of becoming the new natural theology. Science has come to understand the working of nature so precisely, they argue, that this principle now justifies, explains, and, indeed, proves the very existence of a divine origin of the laws of the universe. The appeal of the anthropic principle, particularly to those who have religious leanings and who may not be scientists themselves, may be seductive: to wit, "I showed you; there is a God and science has now proved it."

Science may have metaphysical and value-laden implications, but that is not the role of science qua science. To attach this much weightier aspect to science is, in the final analysis, ill-conceived and dangerous.

### Limitations of Modern Science

The entire SETI program is based on more than a single theory or paradigm, or even collection of theories and paradigms. It is more like an applied science or technology in which a broad range of scientific ideas and methods are brought to bear on a problem of enormous scope. Virtually all the sciences of observational astronomy, astrophysics, and cosmology, as well as the entire range of evolutionary biology, genetics, and chemistry are involved. In addition, specialists in physics—including such fields as relativity, fundamental particles, fusion and nuclear, and thermodynamics—are needed. Instrumentation and communication technologies and computer specialists are all enlisted in this massive program. The fully conceived SETI program raises some very large substantive and methodological concerns and challenges basic assumptions of the scientific enterprise itself.

One of the most important fundamental assumptions for the doing of science is that science is self-correcting. Only those conjectures that can, in principle, be shown to be false are of genuine scientific interest. If some sort of conjecture does not admit scientific examination, then the conjecture cannot, by definition, admit scrutiny. In a word, it is not falsifiable and is therefore not science. For the SETI program to be a genuine *scientific* program, and not simply a program of *religious* faith, there must be a way to falsify its theoretical claims. In other words, what kind of conjectures would be sufficient, if shown to be false, to invalidate the entire SETI effort?

On the one hand, largely because of the immense complexity of the SETI program, SETI advocates reject this line of reasoning as far too simplistic.[41] Still, SETI advocates have noted that scientists have indeed falsified claims of life on the Moon and Mars and in the rest of the solar system. Next, planets will be either found or not found around the nearest stars; then this hypothesis will be falsifiable for ever-increasing distances.[42] While all of these claims may be falsified, the problem with this argument is that this is falsifiability ad infinitum, ad nauseam. On the other hand, there is only one such conjecture that would settle in favor of the pessimists: it must be *demonstrated*, both empirically and theoretically, that the scientific basis—for example, the biochemical or evolutionary theories—invalidates the foundation for the SETI claims.

None of the SETI advocates are willing to accept the counterarguments provided by some biochemists and evolutionists. Although the SETI effort has not yet discovered the existence of extraterrestrial civilizations, SETI advocates simply argue that we must keep looking because eventually we will find them. If we discover only one extra-terrestrial civilization, our conjecture is validated and SETI will be vindicated. The pessimists ask how much time, effort, and money must be expended before we convince ourselves that there are no extrater-restrials. In other words, what is the *scientific* cost in terms of human and financial resources needed for the SETI program, considering there is so much dispute about its very rationale? The stakes are very high indeed. A successful SETI project might inject genuine meaning into a sometimes nihilistic human condition fraught with such enormous challenges as population growth, environmental degradation, and self-destructive national self-interest, providing possibilities well beyond our wildest thoughts. Alternatively, rejecting the program now would save wasted money and human effort, which could be spent more productively elsewhere.[43] In fairness to the advocates, however, this latter argument diminishes considerably, because current estimates suggest that, in today's dollars, SETI will eventually cost perhaps 250 million dollar—an amount that, admittedly large by everyday stan-dards, pales in comparison with the funding of many other, far less potentially rewarding, programs.

Assuming for the moment that there is sufficient *scientific* justifi-cation, the SETI program is in many respects science operating at the frontier of knowledge, but so have many scientific hypotheses, such as the heliocentric cosmology of Copernicus and, more recently, both the special and general theories of relativity. Whatever the case, SETI advocates are fiercely dedicated to the work of vindicating the bio-physical cosmology—a cosmology that may become the worldview of the twenty-first century.

## Mormonism and SETI

Although various forms of the anthropic principle can have been used to argue in favor of the biophysical cosmology and SETI, each form compels either an epistemological and scientific relativism or a natural theology. While the urge to use these arguments may be se-ductive, the dangers in so doing may leave us with a science, and a religion, that is ultimately without a secure foundation. If we refrain from our consideration of the larger biophysical cosmology, however, there still remains a rather severe criticism of the SETI position put forward by the biologists.

Although possibilities are always open for the existence of extra-terrestrials (after all, it requires only one contact to settle the question), and despite the vocal presence of some advocates, such prominent evolutionists as Ernst Mayr and George Gaylord Simpson consider the biological argument particularly devastating. This places Mormons, and indeed many Christians, in a potentially serious position. On the one hand, there are some really powerful and almost persuasive scientific arguments against the SETI position. On the other hand, in Mormon theology God (male/female) propagates spiritual and eventually physical progeny that will ultimately require worlds without number. Consequently, Mormons are scripturally and theologically committed to the idea of the existence of extraterrestrials and advanced civilizations throughout the universe.

The seduction of a natural theology could divert the concerns of Mormons and lead them to divest themselves of this portion of their theology—a highly unlikely move, because astronomical pluralism is intimately wedded to the entire fabric of Mormon beliefs. Conversely, Mormons could suspend final judgment on the SETI question and only tentatively accept the results (and speculations) of science without taking a hardline position either way. Although somewhat ambiguous, the latter has the virtue of not following the historic course of natural theology to jettison religious convictions. The natural theology approach is both useless and unproductive, and arguments against SETI, though far from conclusive, are severe enough that they should be taken seriously.

But what of either space travel or radio communication? The many theophanies of Joseph Smith must clearly rank as his central temporal experiences. The First Vision of 1820, in which he directly experienced God and Christ, the visits of Moroni from 1823 to 1827, the encounters with John the Baptist and Peter, James, and John between 1829 and 1831, the 1836 experience in the Kirkland Temple, as well as many additional visitations during the intervening years until his death profoundly shaped Joseph Smith's understanding of God, man, and nature. In all these theophanies Joseph Smith related that these "messengers" (or "angels," a term rarely used in Mormon discussion because of the traditional Christian connotation denied by Mormons) were actually resurrected beings who had lived in some earlier mortal condition and now were embodied with a tangible, but immortal, body. Assuming that Moroni actually resides on another world, in secular terminology these "angels" are the Mormon equivalent of extraterrestrials. For Mormon extraterrestrials to visit the earth, they must have an understanding of physical laws far in advance of that of twentieth-century science. Therefore, space travel, though it may be impossible as humans normally think of it, may be a simple matter for those with such understanding.

Considering the interest SETI advocates have been able to generate among the public at large and within certain scientific institutions, given the current state of the SETI debate, Mormonism, which is theologically committed to the doctrine of 'worlds without number,' has at least the following options: (1) to reject SETI on scientific grounds as an ill-conceived, scientifically immature program, (2) given its extreme complexity, to consider SETI largely an exercise in scientific speculation, or (3) to emphasize Mormonism's historic view of the primacy of its revelatory epistemology over strictly empiricological ways of knowing. To embrace science uncritically (e.g., Roberts, Widtsoe) may ultimately require Mormonism to face a theological dilemma.

Although various suggestions might be offered to explain the complete lack of verifiable contact with extraterrestrials, in the final analysis all suggestions must remain speculative. Still, one would think that somewhere in the vastness of interstellar space there exists at least one planet inhabited by *pre*-resurrected, *mortal* beings who have reached the scientific/technological stage from which their electromagnetic seepage would be detectable. The fact remains, however, that presently the (electromagnetic) heavens are quiet—utterly quiet.

## NOTES

1. The phrase 'biophysical cosmology' was first introduced in Steven J. Dick's "Concept of Extraterrestrial Intelligence—An Emerging Cosmology," *Planetary Report* 9, no. 2 (March/April 1989), 13–17.

2. For an overview of the SETI arguments, see Alfred Adler, "Behold the Stars," 224–27; and David Schwartzmann, "The Absence of Extraterrestrials on Earth and the Prospects for SETI," 264–66, both in *The Quest for Extraterrestrial Life: A Book of Readings*, ed. Donald Goldsmith (Mill Valley, Calif.: University Science Books, 1980).

3. Easily the best attempt to understand the difficulty of this problem is John Sorenson, *An Ancient American Setting for the Book of Mormon* (Salt Lake City: Deseret Book, 1985).

4. For details of the Mars controversy, see William G. Hoyt, *Lowell and Mars* (Tucson: University of Arizona Press, 1976), and Michael J. Crowe, *The Extraterrestrial Life Debate, 1750–1900: The Idea of a Plurality of Worlds from Kant to Lowell* (New York: Cambridge University Press, 1986), 480–546.

5. Crowe, *The Extraterrestrial Life Debate*, 545. Both the Waterfield and the Sagan quotes are also found in ibid.

6. In 1984 the International Astronomical Union sponsored its 112th symposium, which was dedicated to the most recent developments in SETI. Among the invited participants were the world's leading authorities. See Michael D. Papagiannis, ed., *The Search for Extraterrestrial Life: Recent Developments* (Dordrecht: D. Reidel, 1985).

7. See Karl S. Guthke, *The Last Frontier: Imaging Other Worlds from the Copernican Revolution to Modern Science Fiction*, trans. Helen Atkins (Ithaca: Cornell University Press, 1990).

8. This formulation of the equation was first introduced by Frank Drake in November 1961, during an informal meeting of astronomers to discuss the SETI question. I am using a more recent formulation of the equation modified by Drake and Carl Sagan. Also see Robert E. Machol, "An Ear to the Universe," in Goldsmith, *The Quest for Extraterrestrial Life*, 155; and Reinhard Breuer, *Contact with the Stars: The Search for Extraterrestrial Life* (San Francisco: Freeman, 1982), 87–90.

9. While we are limiting this discussion to the Milky Way Galaxy (classified as a spiral galaxy), within the universe as a whole about 80 percent of all galaxies are classified as elliptical. Formed in the earliest epoch of the universe, these galaxies contain very old stars, primarily composed of only hydrogen and helium, and therefore utterly useless for life, because life apparently requires the heavier elements. Consequently, the arguments which follow would apply to approximately only 20 percent of all galaxies.

10. Until recently most astronomers rejected the reliability of the meager evidence for extrasolar nonluminous planets. At the time of this writing, a British astronomical team led by Andrew G. Lyne announced in *Nature* (25 July 1991) the discovery of such a planet. See R. Cowen, "Wavering Radio Signals Hint at an Unseen Planet Orbiting a Pulsar," *Science News* 140, no. 4 (27 July 1991), 53. One of the crucial projects for which the Hubble Space Telescope was designed is to search for such evidence; see Robert W. Smith, *The Space Telescope: A Study of NASA, Science, Technology, and Politics* (New York: Cambridge University Press, 1989).

11. Bernard M. Oliver has argued that "stable orbits close to each component [star] are certainly possible" ("Proximity of Galactic Civilizations," in Goldsmith, *The Quest for Extraterrestrial Life*, 181).

12. I have chosen probability estimates from a variety of sources giving reasonably conservative numbers that reflect the range of values among SETI proponents and critics. See Carl Sagan, "Direct Contact among Galactic Civilizations by Relativistic Interstellar Spaceflight," in Goldsmith, *The Quest for Extraterrestrial Life*, 205–13; Nicholas Rescher, "Extraterrestrial Epistemology," in *Extraterrestrials: Science and Alien Intelligence*, ed. Edward Regis, Jr. (New York: Cambridge University Press, 1987), 83–116; and Breuer, *Contact with the Stars*, 87–90, 212–13.

13. Hollis R. Johnson, "Civilizations Out in Space," *Brigham Young University Studies* 11, no. 1 (Autumn 1970), 6, 9.

14. The following discussion relies heavily on the arguments put forward by Sagan, "Direct Contact among Galactic Civilizations," 206–8; also see Johnson, "Civilizations Out in Space," 6–9.

15. Although other possibilities may come to mind, such as ones suggested by religious, mystical or quasi-scientific communities, because this is a discussion of religion and *science*, I am considering only scientific possibilities. For problems entailed in the migration of humans into space, see *Interstellar*

*Migration and the Human Experience,* ed. Ben R. Finney and Eric M. Jones (Berkeley: University of California Press, 1985); and Harry L. Shipman, *Humans in Space: Twenty-First Century Frontiers* (New York: Plenum Press, 1989).

16. See Breuer, *Contact with the Stars,* 241–54. Also see Edward Purcell, "Radioastronomy and Communication through Space," 192–93; Sebastian von Hoerner, "The General Limits of Space Travel," 197–204; and Carl Sagan, "Direct Contact among Galactic Civilizations," 208–10, all in Goldsmith, *The Quest for Extraterrestrial Life.*

17. Purcell, "Radioastronomy and Communication through Space," 193.

18. Sagan, "Direct Contact among Galactic Civilizations," 208.

19. See Frank Drake, "How Can We Detect Radio Transmission from Distant Planetary Systems?," 114–17; idem, "Project Ozma," 118–21; Philip Morrison, "Interstellar Communication," 122–31; Alastair G. W. Cameron, "Communicating with Intelligent Life on Other Worlds," 132–35; Edward Purcell, "Radioastronomy and Communication through Space," 188–92, 194–96; and Sagan, "Direct Contact among Galactic Civilizations," 211–12, all in Goldsmith, *The Quest for Extraterrestrial Life;* and Breuer, *Contact with the Stars,* 91–189.

20. I am indebted to David H. Bailey, a computer scientist at NASA–Ames Research Center, for bringing this crucial problem to my attention. Bailey and a colleague have recently succeeded in developing a solution, dubbed the "fractional Fourier transform," that allows for the detection of these extraterrestrial signals.

21. See Thomas R. McDonough, *The Search for Extraterrestrial Intelligence: Listening for Life in the Cosmos* (New York: John Wiley, 1987), 161–71, 219–21, and Paul Horowitz, John Forster, and Ivan Linscott, "The Eight-Million Channel Narrowband Analyzer," in *The Search for Extraterrestrial Life,* ed. Papagiannis, 361–71.

22. The question concerning the whereabouts of extraterrestrials was first posed by Italian physicist Enrico Fermi in 1950; see McDonough, *The Search for Extraterrestrial Intelligence,* 191.

23. Michael Hart, "An Explanation for the Absence of Extraterrestrials on Earth," 228–31; Laurence Cox, "An Explanation for the Absence of Extraterrestrials on Earth," 232–35; and Sebastian von Hoerner, "Where Is Everybody?" 250–54, all in Goldsmith, *The Quest for Extraterrestrial Life.*

24. John Ball, "The Zoo Hypothesis," 241–42; Michael Papagiannis, "Are We All Alone, or Could They Be in the Asteroid Belt?" 243–45; and David Stephenson, "Extraterrestrial Cultures within the Solar System," 246–49, all in Goldsmith, *The Quest for Extraterrestrial Life.*

25. Frank J. Tipler, "We Are Alone," *Discover* (March 1983), 56, 60–1; and idem, "Extraterrestrial Intelligent Beings Do Not Exist," in *Extraterrestrials: Science and Alien Intelligence,* ed. Edward Regis, Jr. (New York: Cambridge University Press, 1987), 133–50.

26. For a rebuttal to this argument, see Jill Tarter, "Planned Observational Strategy for NASA's First Systematic Search for Extraterrestrial Intelli-

gence (SETI)," in Finney and Jones, *Interstellar Migration and the Human Experience*, 314–30.

27. See Hubert Dreyfus, *What Computers Can't Do: The Limits of Artificial Intelligence* (New York: Harper and Row, 1979), and Roger Penrose, *The Emperor's New Mind: Concerning Computers, Minds, and the Laws of Physics* (New York: Oxford University Press, 1989).

28. William Markowitz, "The Physics and Metaphysics of Unidentified Flying Objects," 255–61; and David Stephenson, "Extraterrestrial Intelligence," 262–63, both in Goldsmith, *The Quest for Extraterrestrial Life*.

29. Though suffering from a polemical cast, one of the best and most representative critical treatments of the UFO hypothesis given by Robert Sheaffer, *The UFO Verdict: Examining the Evidence* (Buffalo: Prometheus Books, 1986). Perhaps the finest critical, nonpolemical discussion of the entire UFO phenomenon is given by David Michael Jacobs, *The UFO Controversy in America* (Bloomington: Indiana University Press, 1975). Among reputable scholars and scientists, the Society for Scientific Exploration is easily the most responsible organization which explores UFO and other anomalous phenomena. In the *Journal of Scientific Exploration*, a publication of the society, see the most recent reliable research and discussion of UFOs: Jacques Vallee, "Five Arguments against the Extraterrestrial Origin of Unidentified Flying Objects," 4, no. 1 (1990), 105–20; Robert M. Wood, "The Extraterrestrial Hypothesis Is Not That Bad," 5, no. 1 (1991), 103–11; and Jacques Vallee, "Toward a Second-Degree Extraterrestrial Theory of UFOs: A Response to Dr. Wood and Prof. Bozhick," 5, no. 1 (1991), 113–20. For a reliable assessment of the proto-god hypothesis, see Ronald Story, *Guardians of the Universe?* (New York: St. Martin's Press, 1980).

30. George Gaylord Simpson, "The Nonprevalence of Humanoids," 214–21; Sidney Fox, "Humanoids and Proteinoids," 222–23, both in Goldsmith, *The Quest for Extraterrestrial Life*, and Breuer, *Contact with the Stars*, 38–62.

31. Michael Hart, "Habitable Zones about Main Sequence Stars," in Goldsmith, *The Quest for Extraterrestrial Life*, 236–40.

32. For a thorough discussion of the 'anthropic principle,' see John D. Barrow and Frank J. Tipler, *The Anthropic Cosmological Principle* (Oxford: Oxford University Press, 1986), and Errol E. Harris, *Cosmos and Anthropos: A Philosophical Interpretation of the Anthropic Cosmological Principle* (New York: Humanities Press, 1991). I am indebted to Professor Philip Grier, my colleague and a philosopher at Dickinson College, for providing me prepublication access to Professor Harris's invaluable manuscript.

33. Barrow and Tipler, *Anthropic Cosmological Principle*, 16.

34. Robert Jastrow, *God and the Astronomers* (New York: Warner, 1980).

35. See Barrow and Tipler, *Anthropic Cosmological Principle*, 21.

36. Harris, *Cosmos and Anthropos*, 5, 4, 6, 6, respectively.

37. Barrow and Tipler, *Anthropic Cosmological Principle*, 21–22. Harris has argued in his *Cosmos and Anthropos* that Wheeler's 'participatory anthropic principle' is incoherent (see 9–12).

38. Two of the most celebrated examples of world-renowned modern scientists discussing the anthropic principle are Paul Davies, *The Cosmic Blueprint: New Discoveries in Nature's Creative Ability to Order the Universe* (New York: Simon and Schuster, 1988), 152–64, 197–203, and Stephen Hawking, *A Brief History of Time: From the Big Bang to Black Holes* (New York: Bantam Books, 1988), 124–26.

39. For a discussion of the intellectual cleavage that has increasingly separated science from nonscience, see C. P. Snow, *The Two Cultures* (New York: Cambridge University Press, 1959).

40. David Ehrenfeld, *The Arrogance of Humanism* (New York: Oxford University Press, 1978).

41. For one of the clearest constructive critiques of the entire SETI program, see *Extraterrestrials: Science and Alien Intelligence*, ed. Edward Regis, Jr. (New York: Cambridge University Press, 1987); for a devastating critique, see Barrow and Tipler's "The Space-Travel Argument against the Existence of Extraterrestrial Intelligent Life," in their *Anthropic Cosmological Principle*, 576–612.

42. See Dick, "The Concept of Extraterrestrial Intelligence," 13–17.

# CONCLUSION

# Mormonism and Science in the Modern World

This book is based on a variety of themes. Philosophically, it argues that (1) science is a human enterprise, (2) science is forever changing, and (3) the conceptual world of science does not represent necessarily an objective reality but rather an *explanation* of that reality. Although historically science and religion have addressed ostensibly similar issues, (1) science and religion make radically different yet fundamental methodological and metaphysical assumptions without which it would be impossible to do either science or religion; (2) natural theology is a dangerous extrapolation of a scientific "realism"; and (3) there exists no inherent latent or explicit warfare between science and religion. In light of the historic Mormon–science connection, Mormonism has adopted, with only a few exceptions, (1) a very positive view toward science, (2) an epistemological realism canonized by modern scripture, and (3) the possibilities for a retreat from scientific realism.

As we have seen, these various themes are deeply interrelated both historically and conceptually. Whether dealing with Joseph Smith's cosmology; Orson Pratt's astronomy; Roberts, Talmage, Widtsoe, or Merrill's science; neoliteralist theology and anti-evolutionary cosmogony; or the biophysical cosmos, a rigid distinction between science, religion, philosophy, and history is profoundly inadequate. As many of the issues treated in science and religion illustrate, science is no more a descriptive, neutral, and sanitized process than religion. Both represent highly complex cosmologies with profoundly important philosophical views—and assumptions—embedded in a historical matrix.

The Mormon authorities B. H. Roberts, James E. Talmage, John A. Widtsoe, and Joseph F. Merrill, who were very positive about science otherwise, ultimately rejected science in any guise as the foundation for religion or moral science in general. Thus, although some Mormons came close to developing a natural theology, Roberts for example recognized that science by its very nature is epistemologically insecure and therefore metaphysically suspect. Still, Roberts, Widtsoe, and others did come close to making the classic error of perpetuating nineteenth-century Protestant natural theology. Fortunately for Mormonism, they refused ultimately to follow science at the expense of their religious convictions.

A more careful and less polemical discussion of Joseph Smith's ideas, however, would have saved Roberts, Widtsoe, and other Mormons from making problematic and erroneous, but "faith promoting," statements, such as that the ether of nineteenth-century physics is the physical embodiment of the Holy Ghost or that astronomy verifies the truth of the book of Abraham. In their defense, however, many non-Mormons—scientists and others—even to this day have made similar claims about the alleged truthfulness of science. Unwittingly, these individuals were laying the foundation for a full-fledged natural theology: the view that religion can be verified by association with the prevailing theories and findings of science. Although scientific research traditions have great longevity and even theories within those traditions may achieve deep concensus for long periods of time, in fact science is in a continual state of flux. Except for the most absurdly ridiculous (as presently conceived) "scientific" claims, such as the earth is flat or the sun and all the planets revolve about the earth or organisms spontaneously generate from decaying and dead matter, it is utterly presumptuous to imply that religious claims are true or false based on the findings of contemporary natural science. At the deepest, ontological level, science as philosophy is unable to reveal much at all. As the cosmologist Edward R. Harrison has expressed it: "The history of cosmology [and science generally] shows us that in every age devout people believe that they have at last discovered the true nature of the Universe, whereas in each case they have devised a world picture—merely a universe—that is like a mask fitted on the face of the still unknown Universe."[1]

If we avoid philosophical considerations, then we too easily fall into the trap of an either-or, true-false mind-set. For instance, as we have explored, while evolution remains mostly neutral on the divine premise, one version of the anthropic principle argues for possibilities that can be construed as favoring a providential cosmos. Although SETI is equally complex, within a Mormon context it has the potential for a catastrophic negative influence. But drawing these or similar conclusions from any number of complex scientific positions assumes that science really says something about some ultimate, objective reality. Although we could probably agree on the *existence* of an ultimate reality, actually coming to *know* that reality—let alone in some objective way—is a profoundly more difficult, if not impossible, task.

Because of Mormonism's emphasis on a direct correspondence between mind (the perceiver) and body (the perceived), the temptation either to develop a natural theology or minimally to confuse scientific findings with aspects of religion becomes almost seductive. Although some Mormons have tied science too closely to their religion, Mormonism in general never committed the fatal error of developing a natural theology from which it might have been difficult to extricate itself. Within Mormonism, although science has been described sometimes in critical, though usually positive, terms, it has usually been embraced enthusiastically.

Mormonism is a complex movement encompassing a broad ideological spectrum from historical neoliteralism to positive scientism to total secular accommodationism. Although the breadth of these attitudes is neither fully original nor unique within Mormonism, the debate is grounded fundamentally in the relative epistemological status of revelation vis-à-vis naturalism and rationalism. For most of the nineteenth century Mormonism's liberal, post-Enlightenment interpretation of humans and nature compelled it to understand science in a positive, though perhaps naive, way. This view was given canonical status in the Thirteenth Article of Faith, which implies that all truth (presumably including scientific) is subsumed in the restoration process of Mormonism. By the beginning years of the twentieth century, Mormons had not only developed a complex theology but also had acquired enough experience with science and the time to reflect extensively on questions that sprang from nearly a century of contemplation.

Mormonism as a religion has been tentatively accepting of most scientific ideas, with the possible exception of organic evolution. Thus, during the early decades of the twentieth century, after Roberts became a member of the First Quorum of the Seventy and Talmage, Widtsoe, Merrill, and Lyman became members of the Council of the Twelve, they produced a large number of church materials—books, priesthood and mutual manuals, articles, and addresses—that relied heavily on contemporary scholarly thinking and presented scientific matters in a generally positive manner.

Mormonism also consciously presents itself as a revealed religion and therefore supports a revelatory epistemology that emphasizes the knowability of transcendent reality. God, being the author of that reality, is equally the author of the physical universe. That premise urges Mormonism to adopt a positive scientism in which science is understood as *philosophy*—as truth. During the later years of Widtsoe and Merrill's apostolic tenure, however, science as *technology* became tarnished with a slight negativism.

Like most believing religionists, Mormons who are also scientists are therefore caught in an epistemological dilemma. On the one hand, their scientific training has convinced most of the validity of empirical and quantitative processes. On the other hand, the Mormon religious tradition provides a powerful matrix of scriptural evidences, extensive personal religious experiences, and a living prophet, all of which subsumes an extraphysical knowledge context. Given this larger religious vision, Mormons have consistently claimed that traditional epistemological approaches to understanding reality are inadequate.

A consistent understanding of this distinction by many Mormon thinkers, such as Orson Pratt in the early years of the Mormon church and Henry Eyring in our time, should compel Mormons, first, to divest themselves of the obligations of a natural theology. There is simply no compelling theological reason within Mormonism to engage in any form of natural theology. Properly conceived, science is not, and should never become, an intellectual partner of theology—including Mormon theology. Looking at the same concern from the religious side, one can say that genuine faith can *only* be sustained outside the dimensions of historical and scientific evidence.

In so doing, however, Mormons will need to relinquish their realist view of the world that traditionally has supported their understanding of the physical universe (i.e., science). By rejecting Isis, however, they should not be seduced by Osiris: that *no* scientific claims are valid and that all of science is an ill-conceived enterprise. To adopt the latter view is to assent to an irrationalism where any number of ad hoc claims are allowed, because there is no criteria of assessment other than rank prejudice, fear of challenge, or a dogmatism of authority. Although a middle ground is less philosophically secure, it is nevertheless sustained by the history of science that provides incontrovertible evidence that science must be seen in tentative and approximate terms. In short, at its very core, science is not a body of answers; rather, it is a way of asking questions.

As I have argued, natural theology is inherently dangerous for the faithful as well as for the skeptic. But in human terms there is a much greater danger to both religionists and scientists from those who un-critically adopt a scientific realism. Given the view that science some-how actually informs us of the real workings of nature and that science carefully constructed provides an understanding (even if only a glimpse) of the ontology of nature, we find the crucial step to an *ex-clusive* reliance on human abilities.

An understanding of this premise, whether by Mormons or by those in other religious traditions, leads to a set of assumptions that, ironi-cally, undercuts the very foundations of most normative religious views. David Ehrenfeld has labeled those assumptions 'humanist.'[2] These assumptions are: (1) all problems are soluble; (2) all problems are soluble by science, technology, and social engineering; (3) humans are uniquely capable of providing these solutions; (4) some resources are infinite and those that are not have substitutes; (5) as a result, humans can make over nature entirely if necessary; and (6) human civilization will survive!

Although this study has very little to do with the validity of Mormon religious claims, on the science side generally Mormons uncritically continue to believe that true religion and true science will never conflict. Such a foundation for a scientific realism compels the view that humans cannot only know ultimates (scientific though not necessarily moral nor spiritual) but that humans can control, shape, modify, and, in short, totally make over nature itself, if necessary. In the Mormon tradition this view is deeply reinforced by the idea that, except in a few mirac-ulous instances, God and nature are fundamentally governed by nat-ural law. Though Mormonism possesses powerful tendencies toward this epistemological realism, to its credit, when faced with the choice Mormonism has historically deferred to the view that nature is not an object to manipulate and exploit but an entity to understand and revere. In doing so, however, it has not rejected its reliance on realism. Rather, it has drawn deeply from that part of its tradition that compels the view that humans are caretakers of Nature. After all, God created the earth, not for human indulgence and consumption but for the salvation of his creation. Ultimately, therefore, Mormonism's emphasis on its metaphysics and eschatology, and not on that part of its epistemology that encourages a realism, protects it from the arrogance of secular humanism.

## NOTES

1. Edward R. Harrison, *Cosmology: The Science of the Universe* (New York: Cambridge University Press, 1981), 1.

2. David Ehrenfeld, *The Arrogance of Humanism* (New York: Oxford University Press, 1978).

and God said:

$$t' = \frac{t - vx/c^2}{\sqrt{1 - v^2/c^2}}$$

$$m = \frac{m_0}{\sqrt{1 - v^2/c^2}}$$

$$u(\lambda) = \frac{8\pi hc\lambda^{-5}}{e^{hc/\lambda hT} - 1}$$

$$eV_0 = hv - \phi$$

$$\frac{dn}{dt} = n_1 c_2 ... c_m K \frac{kT}{h} e^{-(\Delta F^{++}/RT)}$$

$$c[\alpha p + \beta\mu c]\psi = -\frac{\hbar}{i}\frac{\partial\psi}{\partial t}$$

$$-\frac{\hbar^2}{2m}\frac{\partial^2\psi}{\partial x^2} + V(x)\psi = i\hbar\frac{\partial^2\psi}{\partial t^2}$$

$$\Delta x \Delta p \geq \frac{\hbar}{2\pi}$$

and there was

light?

# Glossary

The following is provided to assist the reader not already familiar with some of the terms used in this study. Other technical terms not appearing here are defined within the chapters themselves.

A priori: Independent of experience; thus, a priori claims may be stated prior to any experiential or empirical knowledge.

Anthropocentrism: The view that human beings are the most important entity in the universe.

Anthropomorphism: The conception of God in human terms with human-like characteristics.

Contingency: The state of that which exists but which could possibly not exist; opposed to necessity.

Cosmology: The theory of the structure of reality by means of either scientific investigation or metaphysical speculation.

Cosmogony: The theory of the origins of reality by means of either scientific investigation or metaphysical speculation.

Creationism: In science the view that the entire natural world is the product of a special and direct creative act of God. Thus, in the biological world each nonhuman species, and among humans each individual soul, is the result of a special and direct creative act of God.

Epistemology: The theory of the nature of knowledge—how we know what we know.

Eschatology: The doctrine of the end of the history of the world.

Hermeneutics: Methodological principles of interpretation (as of the scriptures).

Metaphysics: The discipline that deals with the nature of reality, as distinguished from the natural sciences (physics) by its nonempirical and speculative basis.

Necessity: The state of that which exists and which could not possibly not exist; opposed to contingency.

Ontology: The discipline that deals with the nature of being.

Realism. In epistemology, the view that the object of knowledge is real, independent of its being known. In science, the view that scientific knowledge is an accurate representation of what is. Thus in 'epistemological realism' that which is real is defined in terms of what is knowable.

# Bibliographic Essay

## Science and Religion

The classic treatment of issues in science and religion that explores the nature of each discipline and surveys the relationship between them both topically and historically is given by Ian Barbour, *Issues in Science and Religion* (New York: Harper and Row, 1966). Easily the finest scholarly examination to date that unravels the complex relation historically between science and religion on topics from the Middle Ages into the twentieth century is provided by David C. Lindberg and Ronald L. Numbers, eds., *God and Nature: Historical Essays on the Encounter between Christianity and Science* (Berkeley: University of California Press, 1986). A re-examination of the warfare thesis is given in David C. Lindberg and Ronald L. Numbers, "Beyond War and Peace: A Reappraisal of the Encounter of Christianity and Science," *Church History* 55 (1986), 338–54. The best exploration of historiographic issues dealing exclusively with the American milieu is found in Ronald L. Numbers, "Science and Religion," *OSIRIS: A Research Journal Devoted to the History of Science and Its Cultural Influences* 1 (1985), 59–80.

The clearest, best, and most accurate account of the Galileo affair and the Catholic church is found in Jerome Langford, *Galileo, Science and the Church* (Ann Arbor: University of Michigan Press, 1971), where the author argues that the alleged conflict between Galileo and the church was more a conflict of personality between an arrogant Galileo and a overly reactionary Cardinal Bellarmine; for a complementary argument, see Stillman Drake, *Galileo* (New York: Oxford University Press, 1980).

The cultural influences of natural theology are explored in several excellent monographs. Richard S. Westfall, in his essential *Science and Religion in Seventeenth-Century England* (Ann Arbor: University of Michigan Press, 1973), examines the role of the leading scientific virtuosi and their impact on the emergence of natural theology and argues that the new science of the seventeenth century laid the groundwork for the emergence of a scientific naturalism in the eighteenth century. Charles C. Gillispie, in *Genesis and Geology: The Impact of Scientific Discussion upon Religious Beliefs in the Decades before Darwin* (New York: Harper and Row, 1959), argues that the alleged conflict between science and religion was a function more of religion *in* science rather than religion *versus* science, while Herbert Hovenkamp, in *Science and Religion in America, 1800–1860* (Philadelphia: University of Pennsylvania Press, 1978), details the decline of natural theology and the eventual problems it entailed for religion in the American milieu. For a detailed review of the latter, see Erich Robert Paul, "Natural Theology," *Dialogue* 12, no. 2 (Summer 1979), 134–37.

Two important studies have explored the view that Christian culture laid the foundations for the emergence of modern science. In *Religion and the Rise of Modern Science* (Grand Rapids: Eerdman's, 1972), Reijer Hooykaas argues that modern science is a product of a biblical worldview, while Eugene M. Klaaren, *Religious Origins of Modern Science* (Grand Rapids: Eerdman's, 1977), argues that modern science is a result primarily of a theology of creation.

The relationship of Anglo-American culture and Protestantism is explored in several excellent studies. Frank M. Turner, in *Between Science and Religion: The Reaction to Scientific Naturalism in Late Victorian England* (New Haven: Yale University Press, 1974), examines the rejection of the dogmas of both science and religion by several leading British intellectuals. Pietro Corsi, in *Science and Religion: Baden Powell and the Anglican Debate, 1800–1860* (New York: Cambridge University Press, 1988), explores the role that Baden Powell played in the discussion of natural theology during the early decades of the nineteenth century. John Dillenberger, in *Protestant Thought and Natural Science: A Historical Interpretation* (New York: Doubleday, 1960), analyzes the role that the Protestant ethos played in the nascent stages of modern science. Theodore Dwight Bozeman, in his *Protestants in an Age of Science: The Baconian Ideal and Antebellum American Religious Thought* (Chapel Hill: University of North Carolina Press, 1977), explores the role of the Protestant ethos in the first half of the nineteenth century. Donald Fleming, in his biography *John William Draper and the Religion of Science* (New York: Octagon, 1972), assesses the role that John W.

Draper played in antireligious attitudes in nineteenth-century America. Paul A. Carter, in his provocative study *The Spiritual Crisis of the Gilded Age* (DeKalb: Northern Illinois University Press, 1971), explores the impact of naturalism, secularism, and science on religious sensibilities during the last third of the nineteenth century.

Consensus has it that the impact of Darwinism on the cultural thinking of nineteenth- and twentieth-century religious ideas has been profound. The single best source to begin one's understanding of this is James R. Moore, *The Post-Darwinian Controversies: A Study of the Protestant Struggle to Come to Terms with Darwin in Great Britain and America, 1870–1900* (New York: Cambridge University Press, 1979). Peter J. Bowler, in his controversial *Non-Darwinian Revolution: Reinterpreting a Historical Myth* (Baltimore: Johns Hopkins University Press, 1988), denies the traditional interpretation; rather, he suggests that Darwin's theory was only one part of a transition to an evolutionary viewpoint and consequently that it affected religion and social attitudes far less than hitherto thought. Neal C. Gillespie, in *Charles Darwin and the Problem of Creation* (Chicago: University of Chicago Press, 1979), examines the shift from a creationist to a positivist understanding of nature in the thought of Charles Darwin. Extremely provocative essays on the relationship of evolutionary thinking and religion can be found in John Durant, ed., *Darwinism and Divinity: Essays on Evolution and Religious Belief* (New York: Oxford University Press, 1985).

One of the most provocative examinations of a protoscientific religious relationship is found in Giorgio de Santillana and Hertha von Dechend, *Hamlet's Mill: An Essay Investigating the Origins of Human Knowledge and Its Transmission through Myth* (Boston: David R. Godine, 1977), in which they argue that all great myths, which represent preliterate forms of science, have one common origin in a cosmology of celestial dimensions.

### Religion in America

The standard source for a comprehensive examination of American religion in its social context is Sydney E. Ahlstrom, *A Religious History of the American People* (New Haven: Yale University Press, 1972). An examination of the four great religious 'awakenings' in American history is found in William G. McLoughlin, *Revivals, Awakenings, and Reforms: An Essay on Religion and Social Change in America, 1607–1977* (Chicago: University of Chicago Press, 1978). A discussion of American intellectual thought, including religion, is given in Merle Curti, *The Growth of American Thought*, 3d ed. (New York: Harper and

Row, 1964). Millennialist thought is explored in John F. C. Harrison, *The Second Coming: Popular Millenarianism, 1780–1850* (New Brunswick: Rutgers University Press, 1979); in Paul E. Johnson, *A Shopkeeper's Millennialism: Society and Revivals in Rochester, New York, 1815–1837* (New York: Hill and Wang, 1978); and in W. H. Oliver, *Prophets and Millennialists: The Uses of Biblical Prophecy in England from the 1790s to the 1840s* (London: Oxford University Press, 1978). A discussion of evangelical fundamentalism during Mormonism's transitional period is given in George M. Marsden's *Fundamentalism and American Culture: The Shaping of Twentieth-Century Evangelicalism, 1870–1925* (New York: Oxford University Press, 1980).

A variety of related religious themes that have bearing on developments in Mormonism may be found in Herbert Leventhal, *In the Shadow of the Enlightenment: Occultism and Renaissance Science in Eighteenth-Century America* (New York: New York University Press, 1976), who discusses the impact of the magic worldview in the centuries between Isaac Newton and Joseph Smith; in Marion B. Stowell, *Early American Almanacs: The Colonial Weekly Bible* (New York: Burt Franklin, 1977), who explores the impact of almanacs on American cultural and religious history; and in Dorothy Ann Lipson, *Freemasonry in Federalist Connecticut* (Princeton: Princeton University Press, 1977), who examines masonic thought during the years of the early American republic.

## Mormonism in America

Two of the most reliable sources dealing with the history of Mormonism are James B. Allen and Glen Leonard, *The Story of the Latter-day Saints* (Salt Lake City: Deseret Book, 1976), and Leonard Arrington and Davis Bitton, *The Mormon Experience: A History of the Latter-day Saints* (New York: Alfred Knopf, 1979). A careful assessment of Mormonism as an American phenomenon is discussed by various scholars in Thomas G. Alexander and Jessie L. Embry, eds., *After 150 Years: The Latter-day Saints in Sesquicentennial Perspective* (Salt Lake City: Signature, 1983). A discussion of Mormonism's transition into the American mainstream is presented by Thomas G. Alexander in his award-winning *Mormonism in Transition: A History of the Latter-day Saints, 1890–1930* (Urbana: University of Illinois Press, 1986).

Thematic analyses of Mormonism abound. For a provocative examination of some of the dilemmas of the Mormon experience, see Klaus Hansen, *Mormonism and the American Experience* (Chicago: University of Chicago Press, 1981). Perhaps one of the most thought-provoking and original essays in recent years, arguing that Mormonism represents the emergence of a new religious tradition, is given in Jan Shipps, *Mormonism: The Story of a New Religious Tradition* (Urbana: University of Illinois Press, 1985). Although far from conclusive, for an argument that Joseph Smith's world was saturated with a magical substratum into which he delved to construct Mormonism, see D. Michael Quinn, *Early Mormonism and the Magic Worldview* (Salt Lake City: Signature, 1987). For a discussion of Quinn's controversial thesis, see Paul H. Peterson, ed., "Book Reviews," *Brigham Young University Studies* 27, no. 4 (Fall 1987), 87–121.

The biographies of Joseph Smith and the founding of Mormonism are legion. For an examination of the influences on Joseph Smith from his heritage, see Richard L. Anderson, *Joseph Smith's New England Heritage* (Salt Lake City: Deseret Book, 1971). For a careful examination of the historical veracity of Joseph Smith's first theophany, see Milton V. Backman, Jr., *Joseph Smith's First Vision*, 2d ed. (Salt Lake City: Bookcraft, 1980). Perhaps one of the most original and reliable histories of Joseph Smith by a leading American historian is given by Richard L. Bushman, *Joseph Smith and the Beginnings of Mormonism* (Urbana: University of Illinois Press, 1984). A useful biography of Joseph Smith which can be consulted with profit is Donna Hill, *Joseph Smith: The First Mormon* (New York: Doubleday, 1977).

Several biographies of important Mormon thinkers who were interested in science are available. Breck England, in his *Life and Thought of Orson Pratt* (Salt Lake City: University of Utah Press, 1985), discusses Mormonism's first scientist-philosopher. Unfortunately, England fails to assess the philosophical-scientific milieu within which Pratt worked, and thus he does not convey the richness of Pratt's worldview. Truman G. Madsen, in his *B. H. Roberts: Defender of the Faith* (Salt Lake City: Bookcraft, 1980), presents a biography of Mormonism's most articulate defender of the faith roughly during the forty years after the Mormon church entered mainstream American life with the granting of statehood to Utah. In his biography, however, Madsen fails to discuss the two most controversial episodes of Roberts's later career: the circumstances surrounding his writing of "The Truth, the Way, the Life" and his examination of the veracity of the Book of Mormon. Elsewhere, Madsen does discuss the former in "The Meaning of Christ—'The Truth, the Way, the Life': An Analysis of B. H. Roberts' Unpublished Masterwork," *Brigham Young University Studies* 15 (1975), 259–92. On Roberts's Book of Mormon studies, see B. H. Roberts, *Studies of the Book of Mormon*, ed. Brigham D. Madsen (Urbana: University of Illinois Press, 1985). An insightful assessment of Roberts is given in Sterling

M. McMurrin's "B. H. Roberts: Notes on a Mormon Philosopher-Historian," *Dialogue* 2 (Winter 1967), 141–49, reprinted from McMurrin's introduction to B. H. Roberts, *Joseph Smith: The Prophet-Teacher* (Princeton: Deseret Club of Princeton University, 1967). An apologetic biography of James E. Talmage is given in John R. Talmage, *The Talmage Story: Life of James E. Talmage—Educator, Scientist, Apostle* (Salt Lake City: Bookcraft, 1972). Dale C. LeCheminant, in his "John A. Widtsoe: Rational Apologist" (Ph.D. dissertation, University of Utah, 1977), has written a useful, though not always reliable, biography of one of Mormonism's most influential scientists.

For a profoundly important analysis of Mormon theology, see Sterling M. McMurrin, *The Theological Foundations of the Mormon Religion* (Salt Lake City: University of Utah Press, 1965). This book is required reading for anyone who aspires to an understanding of Mormon theology. O. Kendall White, in his important *Mormon Neo-Orthodoxy: A Crisis Theology* (Salt Lake City: Signature, 1987), examines the drift to theological conservatism in recent decades. Thomas Alexander, in an insightful essay "Reconstruction of Mormon Theology: From Joseph Smith to Progressive Theology," *Sunstone* 5, no. 4 (July/August 1980), 24–33, explores changing developments in Mormon theology. Examinations of Mormon intellectual trends are given in Davis Bitton, "Anti-Intellectualism in Mormon History," *Dialogue* 1, no. 3 (Autumn 1966), 111–34, and in Leonard Arrington, "The Intellectual Tradition of the Latter-day Saints," *Dialogue* 4, no. 1 (Spring 1969), 13–26.

## Science and Mormonism

The most comprehensive bibliography of materials dealing with science and Mormonism is given in Robert L. Miller, "Science/Mormonism Bibliography: 1830–1989." The historiography dealing with science, scientific theory, and Mormon beliefs began with Parley P. Pratt and Orson Pratt. In book form, the discussion of the relationship of science and religion began chronologically with John H. Ward's *Gospel Philosophy: Showing the Absurdities of Infidelity and the Harmony of the Gospel with Science and History* (Salt Lake City: Juvenile Instructor, 1884). Nels L. Nelson, in his *Scientific Aspects of Mormonism* (New York: E. P. Dutton, 1904), became the first Mormon author to deal extensively with evolutionary views. First published in serial form in the *Improvement Era* (1903–4), John A. Widtsoe's *Joseph Smith as Scientist* (Salt Lake City: Mutual Improvement Association, 1908) argues that Joseph Smith anticipated many important scientific discoveries. Although not all of Roberts's theological speculations deal with science, B. H. Roberts, in *The Seventy's Course in Theology. Second Year: Outline History of the Dispensations of the Gospel* (Salt Lake City, 1908), deals with many relevant issues. John A. Widtsoe's *Science and the Gospel* (Salt Lake City, 1908) was the 1908–9 Mutual Improvement Association manual. Frederick J. Pack provides a forceful defense of evolutionary thinking in his *Science and Belief in God* (Salt Lake City: Deseret News Press, 1924). Written for the Mutual Improvement Association, John A. Widtsoe's *How Science Contributes to Religion* (Salt Lake City, 1927) assumes that science and Mormonism are compatible, and therefore science adds to our understanding of the creation. B. H. Roberts in his magnum opus, "The Truth, the Way, the Life: An Elementary Treatise on Theology" (unpublished, 1928), deals with numerous scientific and religious issues. John A. Widtsoe in his *In Search of Truth* (Salt Lake City: Deseret Book, 1930) addresses the questions "What is science?" and "How does the church view science?" In a collection of many of his most important articles, John A. Widtsoe in *Evidences and Reconciliations*, 3 vols. (Salt Lake City: Bookcraft, 1943, 1947, 1951), treats numerous scientific issues that premise the view that Mormonism and modern science are complementary. Joseph F. Merrill in his *Truth-Seeker and Mormonism* (Salt Lake City: Deseret Book, 1946) also premises that Mormonism and true science are compatible.

Following the publication of Joseph Fielding Smith's rabidly anti-science *Man: His Origin and Destiny* (see below), Paul A. Green was commissioned by the Mormon church to compile *Science and Your Faith in God* (Salt Lake City: Bookcraft, 1958), which contains chapters by Henry Eyring, Carl J. Christensen, Harvey Fletcher, Franklin S. Harris, Joseph F. Merrill, Frederick J. Pack, and John A. Widtsoe rebutting Smith. Discussion of many of these issues was continued by Mormon scientists in Frank B. Salisbury, *Truth by Reason and by Revelation* (Salt Lake City: Deseret Book, 1965); in *The Faith of a Scientist* (Salt Lake City: Bookcraft, 1967), in which Henry Eyring, easily Mormonism's most famous scientist, relates how his faith and his science are compatible; in Frank B. Salisbury, *The Creation* (Salt Lake City: Deseret Book, 1976), which provides a discussion of the creation of the world using both scientific and scriptural sources; and in William Lee Stokes, *The Creation Scriptures: A Witness for God in the Scientific Age* (Bountiful, Utah: Horizon, 1979), who argues that the Mormon view of creation is fully compatible with normative science.

Most recently a group of Brigham Young University scientists explored their faith in *Science and Religion: Toward a More Useful Dialogue*, Vol. 1, *Background for Man: Preparation of the Earth*, ed. Wilford M. Hess and Raymond T. Matheny, and Vol. 2, *The Appearance of Man: Re-*

*plenishment of the Earth*, ed. W. M. Hess, R. T. Matheny, and D. D. Thayer (Geneva, Ill.: Paladin House, 1979). These volumes contain articles in their authors' fields of specialty providing personal testimonials of the compatibility of science with Mormonism. For a useful review of this collection of essays, see F. S. Harris, Jr., "Archaeology in a Setting of Science and Religion," *Newsletter and Proceedings of the SEHA* 146 (May 1981), 7–9.

The Fall/Winter 1973 issue of *Dialogue* was devoted to relevant science/Mormon issues; see James L. Farmer, ed., "Science and Religion," *Dialogue* 8, nos. 3/4 (Autumn/Winter 1973), 21–143, which contains articles by James L. Farmer, Richard F. Haglund, Jr., Duane E. Jeffery, Hugh Nibley, Edward L. Kimball, Clyde Parker, Brent Miller, William E. Dibble, William L. Stokes, and Benjamin Urrutia.

For a variety of recent articles which explore various aspects of Mormonism and science, see the following: A. Lester Allen, "Science and Theology: A Search for the Uncommon Denominator," *Brigham Young University Studies* 29, no. 3 (Summer 1989), 71–78; David H. Bailey, "Scientific Foundations of Mormon Theology," *Dialogue* 21, no. 2 (Summer 1986), 61–80; James L. Farmer, William S. Bradshaw, and F. Brent Johnson, "The New Biology and Mormon Theology," *Dialogue* 12, no. 4 (Winter 1979), 71–75; Steven H. Heath, "The Reconciliation of Faith and Science: Henry Eyring's Achievement," *Dialogue* 15, no. 3 (Fall 1982), 87–99; F. Kent Nielsen, "The Gospel and the Scientific View: How Earth Came to Be," *Ensign* 10, no. 9 (September 1980), 67–72; Keith E. Norman, "Mormon Cosmology: Can It Survive the Big Bang?" *Sunstone* 10, no. 9 (1985), 19–23; and Dennis Rowley, "Inner Dialogue: James Talmage's Choice of Science as a Career, 1876–1884," *Dialogue* 17, no. 2 (Summer 1984), 112–30.

For modern Mormon advocates of the plurality of worlds and how it relates to cosmology generally, see R. Grant Athay, "Worlds without Number: The Astronomy of Enoch, Abraham, and Moses," *Brigham Young University Studies* 8, no. 3 (Spring 1968), 255–69; Hollis R. Johnson, "Civilizations Out in Space," *Brigham Young University Studies* 11, no. 1 (Autumn 1970), 3–12; R. Grant Athay, "Astrophysics and the Gospel," *New Era* 2 (September 1972), 14–19; and D. M. McNamara, "The Origin, Structure, and Evolution of the Stars," *Brigham Young University Studies* 8, no. 1 (Autumn 1967), 7–22.

The single best discussion of Mormonism and evolution is given in Duane Jeffrey [*sic*], "Seers, Savants and Evolution: The Uncomfortable Interface," *Dialogue* 8, nos. 3/4 (Autumn/Winter 1973), 41–75; and in Duane Jeffrey [*sic*], "Seers, Savants and Evolution: A Continuing Dialogue," *Dialogue* 9, no. 3 (Autumn 1974), 21–38, which provides responses and a rejoinder to his 1973 article.

Mormon reactions to Darwinian evolution from the late nineteenth century to the present are discussed in Richard Sherlock, "A Turbulent Spectrum: Mormon Reactions to the Darwinist Legacy," *Journal of Mormon History* 5 (1978), 33–59; in Gary James Bergera and Ronald Priddis, "Organic Evolution Controversy," *Brigham Young University: A House of Faith* (Salt Lake City: Signature Books, 1985), 131–71; and in Richard Sherlock, "Campus in Crisis," *Sunstone* 4, no. 1 (January/February 1979), 1016, which treats the "modernist" controversy, focusing largely on the issue of teaching evolution and the use of biblical higher criticism, at Brigham Young University in 1911.

Controversy over the evolution issue among church leaders is discussed in Richard Sherlock, " 'We Can See No Advantage to a Continuation of the Discussion': The Roberts/Smith/Talmage Affair," *Dialogue* 13, no. 3 (Fall 1980), 63–78, which explores B. H. Roberts's views on "pre-Adamites" and the responses by Joseph Fielding Smith and James E. Talmage. Jeffrey E. Keller, in his "Discussion Continued: The Sequel to the Roberts/Smith/Talmage Affair," *Dialogue* 15, no. 1 (Spring 1982), 79–98, continues with an examination of Roberts's views on pre-Adamites and the subsequent responses by Joseph Fielding Smith, Jr., James E. Talmage, and his son Sterling Talmage.

A brief examination of the dating of the geological record and how it accords with Mormonism was recently commissioned by the Mormon church and discussed by Mormon geologist Morris Petersen, "Do we know how the earth's history as indicated from fossils fits with the earth's history as the scriptures present it?" *Ensign* (September 1987), 28–29. His article supports the latest geological findings that indicate that the earth is 4.5 billion years old and maintains that this view does not threaten Mormon beliefs.

### Creationism, Evolution, and Mormonism

Although the tradition extends to the early years of this century, 'creation science' is a recent phenomenon premised on the view that modern science—particularly evolution, geology, and cosmogony—is fundamentally flawed because it does not conform to the "6,000 year" account of a literal Genesis reading. Two authors who argue that Mormonism is compatible with the creation science account of the universe provided most forcefully by evangelical fundamentalist Christianity are Melvin A. Cook, *Science and Mormonism* (Salt Lake City: Deseret News Press, 1967), and Joseph Fielding Smith, *Man: His Origin and Destiny*

(Salt Lake City: Deseret Book Press, 1954). In the latter, the most forceful polemic against normative science from a literal reading of the Mormon canon, Smith argues that Mormonism is compatible only with a creation science account of the universe.

A discussion of the rise of creationism is provided in Ronald Numbers's superb essay "Creationism in Twentieth-Century America," *Science* 218 (1982), 538–44, which appears in expanded form as "The Creationists," in *God and Nature: Historical Essays on the Encounter between Christianity and Science*, ed. D. C. Lindberg and R. L. Numbers (Berkeley: University of California Press, 1986), 391–423.

In sharp contrast, the following books and articles, which may be consulted as a starting place, demonstrate that creation science is neither science nor an account of the creation: Cedric I. Davern, "Evolution and Creation: Two World Views," *Dialogue* 17, no. 1 (Spring 1984), 44–50; and Luther V. Giddings, "Penetrating Muddied Waters: Creationism and Evolution," *Dialogue* 19, no. 1 (Spring 1986), 172–79, both of which provide a devastating critique of creationism and expose the philosophical, epistemological, and scientific inadequacies of creation science. A broader examination of this problem is discussed by Roland M. Frye, ed., in his essential *Is God a Creationist? The Religious Case against Creation Science* (New York: Scribner's, 1983). Frye offers a collection of articles by specialists who argue that normative science—evolution, geology, astronomy—is not only fully compatible with traditional biblical understanding but that creation science is not science and that it impoverishes religion.

The case for evolution opposed to creation science is provided in Douglas J. Futuyma, *Science on Trial: The Case for Evolution* (New York: Pantheon, 1983); and in Laurie R. Godfrey, ed., *Scientists Confront Creationism* (New York: W. W. Norton, 1983), both of whom offer a point-by-point discussion of the validity of normative science vis-à-vis creation science.

## Development of Modern Science

The literature in the history of science has expanded enormously in the last thirty years. For additional bibliography, those interested in serious study of the field should consult *Isis* and *Osiris*, the official journals of the History of Science Society; the *ISIS Cumulative Bibliography*, 13 vols. (London: Mansell, 1971–85), for 1913–65, 1966–75, and 1976–85; and the *Dictionary of Scientific Biography*, ed. Charles Coulston Gillispie and Frederic L. Holmes, 18 vols. (New York: Charles Scribner's Sons).

Though semitechnical, but richly rewarding for the dedicated reader, the most influential book on the history of science published in the last twenty-five years has been Thomas S. Kuhn, *The Structure of Scientific Revolutions*, 2d ed. (Chicago: University of Chicago Press, 1970). Because almost any discussion of the relationship of science and religion must include a substantial discussion of the development of science both historically and philosophically, the following general accounts are suggested. A. M. Alioto, *A History of Western Science* (Englewood, N.J.: Prentice-Hall, 1987), surveys the development of science from antiquity to the present and to its credit relies primarily on contemporary scholarship. Charles C. Gillispie, *The Edge of Objectivity: An Essay in the History of Scientific Ideas* (Princeton: Princeton University Press, 1960), provides an assessment of the emergence of modern science and makes the epistemological premise that science assumes an "objective" reality. David C. Lindberg, ed., *Science in the Middle Ages* (Chicago: University of Chicago Press, 1978), offers the best, most comprehensive, and eminently readable collection of essays by scholars on the topic. Stephen Mason, *A History of the Sciences* (New York: Collier Books, 1962), provides a factual, though dated, survey of science since Greek antiquity. I. Bernard Cohen, in his *Revolution in Science* (Cambridge, Mass.: Harvard University Press, 1985), surveys the development of science using the idea of revolution as a basic organizing principle.

Far and away the most readable and accessible accounts of the development of science from the Middle Ages to the twentieth century are provided by the Cambridge University Press series in the history of science. Edward Grant, in his *Physical Sciences in the Middle Ages* (1971), explores the re-emergence of science from its Greek and Arabic origins. Allen Debus, in *Man and Nature in the Renaissance* (1978), provides a fascinating examination of the sciences and culture during the sixteenth and seventeenth centuries and explores the emergence of magic as a major influence on the rise of modern science. Richard S. Westfall, in *The Construction of Modern Science* (1980), is the best accessible account of the rise and development of modern science in the seventeenth century. Thomas L. Hankins, in *Science and the Enlightenment* (1985), assesses the development of science (mostly French, however) during the eighteenth century. For the nineteenth and twentieth centuries, see P. M. Harman, *Energy, Force and Matter: The Conceptual Development of Nineteenth-Century Physics* (1982); William Coleman, *Biology in the Nineteenth Century* (1971); and Garland Allen, *Life Sciences in the Twentieth Century* (1975).

One of the clearest discussions of the influence of Renaissance naturalism on the emergence of early modern science is given in Francis Yates, *Giordano Bruno and the Hermetic Tradition* (Chicago: University of Chicago Press, 1979). Biographies of some of the most influential scientists of the seventeenth century include Victor E. Thoren, *Tycho Brahe, the Lord of Uraniborg* (New York: Cambridge University Press, 1991), Stillman Drake, *Galileo at Work: His Scientific Biography* (Chicago: University of Chicago Press, 1978), as well as Drake's most recent study, *Galileo: Pioneer Scientist* (Toronto: University of Toronto Press, 1990).

Perhaps the most extensive historical industry in the history of science is that dealing with Isaac Newton. Profitable sources include Gale E. Christianson, *In the Presence of the Creator: Isaac Newton and His Times* (New York: Free Press, 1984), an accessible biography of the greatest of modern scientists. The definitive history of Newton and his intellectual milieu, by one of the world's leading authorities on Newton, is Richard S. Westfall's *Never at Rest: A Biography of Isaac Newton* (New York: Cambridge University Press, 1980). Westfall's earlier *Force in Newton's Physics* (London: Macdonald, 1971) details how Newton arrived at his concept of force and how that idea was used to explain the crucial problems of seventeenth-century science. Betty Jo Teeter Dobbs, in her path-breaking book *The Foundations of Newton's Alchemy* (New York: Cambridge University Press, 1975), shows that Newton's "rational" science was deeply affected by a magical worldview.

Additional works dealing with the emergence of the Scientific Revolution which can be consulted with profit include Richard F. Jones, *Ancients and Moderns: A Study of the Rise of the Scientific Movement in Seventeenth-Century England* (St. Louis: Washington University Press, 1961); Christopher Hill, *Intellectual Origins of the English Revolution* (Oxford: Clarendon Press, 1966); Charles Webster, *The Great Instauration: Science, Medicine, and Reform, 1626–1660* (London: Duckworth, 1975); Hugh Kearney, *Science and Change, 1500–1700* (New York: McGraw-Hill, 1971); I. Bernard Cohen, *The Birth of a New Physics*, rev. ed. (New York: W. W. Norton, 1985); and A. Rupert Hall, *Philosophers at War: The Quarrel between Newton and Leibniz* (New York: Cambridge University Press, 1980).

Among historians of science, the period of the seventeenth century has been the most thoroughly explored, primarily because the roots of modern science are found within a variety of traditions from which the mechanical conception of nature emerged during this period. In addition to the studies by Westfall, Dobbs, Grant, and Debus, one should consult David C. Lindberg and Robert S. Westman, eds., *Reap-*

*praisals of the Scientific Revolution* (New York: Cambridge University Press, 1990), who argue for a revisionist interpretation based on the view that shades of seventeenth-century thinking were present much earlier and threads of the older discredited metaphysics survived until much later. Carolyn Merchant, in her provocative feminist interpretation *Death of Nature: Women, Ecology, and the Scientific Revolution* (San Francisco: Harper and Row, 1980), suggests that the rise of modern science is responsible for the decline of a holistic understanding of nature. For an "externalist," sociological interpretation of the rise of modern science during this period, see I. Bernard Cohen, ed., *Puritanism and the Rise of Modern Science: The Merton Thesis* (New Brunswick: Rutgers University Press, 1990).

For a comprehensive discussion of the historiography of the history of modern physical science, see Stephen G. Brush, *The History of Modern Science: A Guide to the Second Scientific Revolution, 1800–1950* (Ames: Iowa State University Press, 1988), which provides an excellent source briefly outlining the crucial issues in the development of modern science. Thematic histories that may be consulted for value include Stephen G. Brush, *The Temperature of History: Phases of Science and Culture in the Nineteenth Century* (New York: Franklin, 1978), in which he argues that certain major scientific themes of the nineteenth century, such as thermodynamics, were strongly correlated to other significant cultural movements in art and philosophy. Nineteenth-century physics is explored in William Berkson's *Fields of Force: The Development of a World View from Faraday to Einstein* (New York: John Wiley, 1974), and in the demanding Christa Jungnickel and Russell McCormmach, *Intellectual Mastery of Nature: Theoretical Physics from Ohm to Einstein*, 2 vols. (Chicago: University of Chicago Press, 1986). G. N. Cantor and M. J. S. Hodge, *Conceptions of the Ether: Studies in the History of Ether Theories, 1740–1900* (New York: Cambridge University Press, 1981), provide a technical history of etherial studies for the dedicated reader. D. S. L. Cardwell, *From Watt to Clausius: The Rise of Thermodynamics in the Early Industrial Age* (Ithaca: Cornell University Press, 1971), is the standard source examining the emergence of thermodynamics and energy studies.

In addition to the Newton biographies, see A. P. French, ed., *Einstein: A Centenary Volume* (Cambridge, Mass.: Harvard University Press, 1979), a very broad collection of assessable articles on one of the most influential twentieth-century intellects, and Robert Kargon, *The Rise of Robert Millikan: Portrait of a Life in American Science* (Ithaca: Cornell University Press, 1982), who offers a glimpse into the workings of one of America's most influential twentieth-century scientists.

Two of the finest accounts of the rise of technology and its importance in Western thought are David S. Landes, *The Unbound Prometheus: Technological Change and Industrial Development in Western Europe from 1750 to the Present* (New York: Cambridge University Press, 1969), which provides a tour-de-force presentation of the influence of technology on Western society, and Leo Marx, *The Machine in the Garden: Technology and the Pastoral Ideal in America* (New York: Oxford University Press, 1967), which examines the rise and impact of technology in nineteenth-century America.

Equalled only by the Newton industry is that devoted to Darwin. The following may be consulted with profit. Loren Eiseley, *Darwin's Century: Evolution and the Men Who Discovered It* (New York: Doubleday, 1958), a classic and readable, though somewhat dated, account of Darwinian evolution in the nineteenth century. John C. Greene, *The Death of Adam: Evolution and Its Impact on Western Thought* (Ames: Iowa State University Press, 1959), examines the decline of pre-Darwinian thinking and the rise of naturalism. Michael Ruse, *The Darwinian Revolution* (Chicago: University of Chicago Press, 1979), provides a clear and compelling discussion of the rise of Darwinism and its impact on Western thought. The state of Darwinism in America is surveyed, not always reliably, in Cynthia E. Russett, *Darwin in America: The Intellectual Response, 1865–1912* (San Francisco: Freeman, 1976).

A wonderful and highly readable account of the emergence of naturalism during the nineteenth century and its impact on humankind's self-understanding is provided in Loren Eiseley's *The Firmament of Time* (New York: Atheneum Press, 1978). Michael T. Ghiselin, in *The Triumph of the Darwinian Method* (Berkeley: University of California Press, 1969), provides a highly insightful account of the emergence and eventual triumph of the scientific and philosophic implications of Darwin's theory. David L. Hull, *Darwin and His Critics: The Reception of Darwin's Theory of Evolution by the Scientific Community* (Cambridge, Mass.: Harvard University Press, 1974), explores the nature of the Darwinian mindset and examines the influence of and the reaction to the Darwinian worldview. Ernst Mayr, in his *Growth of Biological Thought: Diversity, Evolution, and Inheritance* (Cambridge, Mass.: Belknap Press, 1982), gives a brilliant discussion of biological ideas by one of America's foremost evolutionists. Peter Bowler, in his *Evolution: The History of an Idea* (Berkeley: University of California Press, 1984/89), examines one of the most powerful ideas in Western culture.

On American science, see Brooke Hindle, *Early American Science* (New York: Science History, 1976), which provides an account of the development of science and scientific thinking during the pre-Federal period of American history. Raymond P. Stearns, *Science in the British Colonies of America* (Urbana: University of Illinois Press, 1970), is one of the earliest historians to argue that there was no conflict between science and religion in colonial America. John C. Greene, *American Science in the Age of Jefferson* (Ames: Iowa State University Press, 1984), offers an account of American science during the early Federal period. George Daniels, *American Science in the Age of Jackson* (New York: Columbia University Press, 1968), argues that early nineteenth-century science was above all characterized by a Baconian approach. Robert V. Bruce, *The Launching of Modern American Science, 1846–1876* (New York: Alfred Knopf, 1987), offers an account of American science during its golden era.

## History of Astronomy and Cosmology

A discussion of cosmologies from the ancient Greeks to the seventeenth century is given in Michael J. Crowe, *Theories of the World from Antiquity to the Copernican Revolution* (New York: Dover, 1990). A standard, reliable, and detailed history of astronomy from Greek antiquity to the recent present is given by Anton Pannekoek, *The History of Astronomy* (New York: Barnes and Noble, 1961). Stephen Toulmin and June Goodfield, *The Fabric of the Heavens: The Development of Astronomy and Dynamics* (New York: Harper and Row, 1961), provide a reliable and accessible survey of the history of astronomy from Greek antiquity to the seventeenth century and partially demonstrate the invention of theory and the role of experimental science in theories. The history of astronomy from the later Greek era through the Middle Ages is given by Olaf Pedersen and Mogens Pihl, *Early Physics and Astronomy* (New York: American Elsevier, 1974). A ground-breaking study of the rise of science during the seventeenth century by one of the world's leading authorities is Alexandre Koyre, *From the Closed World to the Infinite Universe* (Baltimore: Johns Hopkins University Press, 1957). A technical examination of the Copernican hypothesis is given in Thomas S. Kuhn, *The Copernican Revolution* (Cambridge, Mass.: Harvard University Press, 1957). The best nontechnical discussion of seventeenth-century planetary astronomy by leading authorities is given in René Taton and Curtis Wilson, eds., *Planetary Astronomy from the Renaissance to the Rise of Astrophysics—Part A: Tycho Brahe to Newton* (New York: Cambridge University Press, 1989).

The development of modern stellar astronomy is provided in a handful of insightful monographs. Michael Hoskin, in his *Stellar Astronomy: Historical Studies* (New York: Science History, 1982), offers perhaps the best thematic account of the development of astronomy and cosmology from Newton to the early years of the nineteenth century. Dieter Herrmann, in his *History of Astronomy from Herschel to Hertz-sprung* (New York: Cambridge University Press, 1984), provides a reliable history with a Marxist orientation. Erich Robert Paul, in *The Milky Way Galaxy and Statistical Cosmology, 1890–1924* (forthcoming from Cambridge University Press) examines the nature of the cosmos during the nineteenth and early twentieth centuries immediately prior to the "Second Astronomical Revolution" of the 1920s and argues that the development of stellar astronomy and cosmology has roots in two parallel scientific traditions. Richard Berendzen, Richard Hart, Daniel Seeley, in their *Man Discovers the Galaxies* (New York: Science History, 1976), offer a semitechnical examination of the events during the 1920s that led to the discovery of the first extragalactic objects. Robert Smith, in his *Expanding Universe: Astronomy's Great Debate, 1900–1931* (New York: Cambridge University Press, 1982), examines one of the history of science's greatest public debates (in 1920) concerning the existence of extragalactic objects. A nontechnical collection of essays by leading authorities on crucial aspects of twentieth-century astronomy is given in Owen Gingerich, ed., *Astrophysics and Twentieth-Century Astronomy to 1950: Part A* (New York: Cambridge University Press, 1984).

Developments in American astronomy are given in Ronald L. Numbers, *Creation by Natural Law: Laplace's Nebular Hypothesis in American Thought* (Seattle: University of Washington Press, 1977), which provides a complete history of ideas about the creation of the cosmos in the nineteenth-century American context.

## Cosmology

There are a variety of reliable and accessible discussions of modern cosmological thinking. These include Benjamin Gal-Or, *Cosmology, Physics and Philosophy* (Berlin: Springer-Verlag, 1981), a reasonably technical text which the dedicated reader will find of value; and Edward R. Harrison, *Cosmology: The Science of the Universe* (New York: Cambridge University Press, 1981), which provides a lay introduction to current thinking about the origin and structure of the cosmos. In his award-winning book, *A Brief History of Time: From the Big Bang to Black Holes* (New York: Simon and Schuster, 1988), Stephen W. Hawking, one of the world's foremost contemporary cosmologists, provides an accessible discussion of the nature of the universe. The nature of the Milky Way Galaxy is explored in Hugo van Woerden and Robert J. Allen, eds., *The Milky Way Galaxy* (Dortrecht, Holland: D. Reidel, 1985), a technical, up-to-date assessment (as of 1985) of our knowledge of our galaxy by professional astronomers.

Some of the cosmic "coincidences" that have been discovered recently by physicists and astronomers are presented in several accessible books. Paul Davies, one of the clearest and most articulate scientists today, explores fundamental issues in science as they pertain to our understanding of humankind and the universe about us in *The Accidental Universe* (New York: Cambridge University Press, 1982). In *God and the New Physics* (New York: Simon and Schuster, 1983), Davies continues the discussion of the nature of a divine universe in light of modern discoveries in fundamental physics. These issues are further examined in Davies's *The Cosmic Blueprint: New Discoveries in Nature's Ability to Order the Universe* (New York: Simon and Schuster, 1988), which emphasizes the 'anthropic' nature of the universe. A tour-de-force treatment of teleological arguments since Greek antiquity, with a critical examination of its modern formulation in the anthropic principle, is given in John D. Barrow and Frank J. Tipler, *The Anthropic Cosmological Principle* (New York: Oxford University Press, 1986). The presentation in this book is comprehensive, relentless, and overwhelming; the reader must be prepared for some deep discussion. Complementing Barrow and Tipler is Errol E. Harris's essential *Cosmos and Anthropos: A Philosophical Interpretation of the Anthropic Cosmological Principle* (New York: Humanities Press, 1991).

Religious and theological issues of contemporary cosmology are discussed in Robert Jastrow's *God and the Astronomers* (New York: Warner Books, 1978), which provides a theological statement by a prominent astronomer who argues that contemporary cosmology has discovered the "genesis" that theologians have suggested for millennia. In Stephen Toulmin, *The Return to Cosmology: Postmodern Science and the Theology of Nature* (Berkeley: University of California Press, 1982), a prominent philosopher examines the relation between science and the proper role of humans in the universe. W. Yourgrau and A. D. Breck, eds., *Cosmology, History, and Theology* (New York: Plenum, 1977), provide a highly provocative discussion by twenty-four leading cosmologists, historians, philosophers, and theologians on the topic of cosmology and religion.

### Plurality of Worlds

By far the two best sources for a discussion of the plurality-of-worlds idea are Steven J. Dick, *Plurality of Worlds: The Origins of the Extraterrestrial Life Debate from Democritus to Kant* (New York: Cambridge University Press, 1982), which traces the extraterrestrial life debate from Greek antiquity through the eighteenth century, and Michael J. Crowe, *The Extraterrestrial Life Debate, 1750–1900: The Idea of a Plurality of Worlds from Kant to Lowell* (New York: Cambridge University Press, 1986), who argues that the extraterrestrial life debate has been shaped by scientific, religious, philosophical, and cultural influences that are deeply interwoven.

The classic, though now dated, discussion of the plurality debate is given by Arthur O. Lovejoy, *The Great Chain of Being* (Cambridge, Mass.: Harvard University Press, 1936), who traces important elements of the plurality-of-worlds debate, among other ideas, from Greek antiquity to the recent present. One of the earliest discussions of this idea was first given in 1794 by Thomas Paine, *The Age of Reason* (Indianapolis: Bobbs-Merrill, 1976), Part I, a highly polemical discussion of the debate written by the foremost defender of deism as the philosophical ground for the American Revolution. Though not always reliable in detail, one of the earliest historical surveys of the scientific idea of multiple inhabited worlds from antiquity to recent times is given by Ralph V. Chamberlin, "Life in Other Worlds: A Study in the History of Opinion," *Bulletin of the University of Utah* 22, no. 3 (1932), 3–52. Although written by a Mormon scientist, it does not treat the concept of pluralism in Mormon theology.

Other sometimes useful sources include Stanley J. Jaki, *Planets and Planetarians: A History of Theories of the Origin of Planetary Systems* (New York: John Wiley, 1977), which provides a comprehensive, though sometimes polemical, discussion of the plurality of worlds debate; and Ivan L. Zabilka, "Nineteenth-Century British and American Perspectives on the Plurality of Worlds: A Consideration of Scientific and Christian Attitudes" (Ph.D. dissertation, University of Kentucky, 1980), a reliable, though narrow, examination of the plurality-of-worlds question in the nineteenth century. David K. Lewis, in *On the Plurality of Worlds* (Oxford: Blackwell, 1986), examines the plurality-of-worlds debate from a philosophical perspective by exploring conceptual "worlds," not actual stellar objects. William G. Hoyt, in his *Lowell and Mars* (Tucson: University of Arizona Press, 1976), offers a biography of Percival Lowell, the principal advocate of the Martian canals theory.

## Search for Extraterrestrial Intelligence

For a general bibliographical source that assesses the modern literature on SETI, see T. B. H. Kuiper and G. D. Brin, "Resource Letter ETC-1: Extraterrestrial Civilization," *American Journal of Physics* 57

(1989), 12–18. Easily the clearest critique of SETI by philosophers, scientists, and SETI specialists is Edward Regis, Jr., ed., *Extraterrestrials: Science and Alien Intelligence* (New York: Cambridge University Press, 1985). There are a number of reliable sources that treat SETI at an accessible level. These include the following: Isaac Asimov, *Extraterrestrial Civilizations* (New York: Fawcett Columbine, 1979); Ronald N. Bracewell, *The Galactic Club: Intelligent Life in Outer Space* (New York: W. W. Norton, 1976); and R. T. Rood and J. S. Trefil, *Are We Alone? The Possibility of Extraterrestrial Civilizations* (New York: Scribner's, 1983).

For a semitechnical discussion of the SETI question for those who have some scientific background, see John F. Baugher, *On Civilized Stars: The Search for Intelligent Life in Outer Space* (Englewood, N.J.: Prentice-Hall, 1985); John Billingham, ed., *Life in the Universe* (Cambridge, Mass.: MIT Press, 1981), and Donald Goldsmith, ed., *The Quest for Extraterrestrial Life: A Book of Readings* (Mill Valley, Calif.: University Science Books, 1980).

Among good, reliable sources, technical discussions of the SETI question by pessimists include Reinhard Breuer, *Contact with the Stars: The Search for Extraterrestrial Life* (San Francisco: Freeman, 1982); and Michael H. Hart and Ben Zucherman, *Extraterrestrials: Where Are They?* (New York: Pergamon Press, 1982).

For a sophisticated and highly technical discussion of the SETI question by leading experts, mostly optimists, from around the world, see the essential Michael D. Papagiannis, ed., *The Search for Extraterrestrial Life: Recent Developments* (Dordrecht, Holland: D. Reidel, 1985). This book contains the results of Symposium no. 112, on the SETI topic, of the International Astronomical Union held at Boston University in 1984.

For the seriously interested reader the entire question of the human expansion into outer space is given in Bruce R. Finney and E. M. Jones, eds., *Interstellar Migration and the Human Experience* (Berkeley: University of California Press, 1986), and in Harry L. Shipman, *Humans in Space: Twenty-First Century Frontiers* (New York: Plenum Press, 1989). Carl Sagan, in his *Dragons of Eden: Speculations on the Evolution of Human Intelligence* (New York: Random House, 1977), provides a highly provocative examination by one of the foremost proponents of SETI of the question of human intelligence as it relates to the SETI debate.

## Philosophy of Science

Philosophy of science attempts to provide an understanding of the nature of science by focusing on such ideas as scientific theories, laws, and basic categories of scientific explanation and by assessing the possibility of growth and development in the scientific enterprise. Developments in modern philosophy of science may be divided into two quite different traditions. The first, known by such names as empiricist and logical-positivist, has attempted to provide a rational reconstruction of science by offering schemes for an ideal explanation of what science ought to be. Rejecting this older tradition as misguided and bankrupt, the second has offered in its place an understanding of science that is built on the historical record, focusing on explaining why science is the way it actually is. The present bibliography reflects the latter approach. John Losee, in his *Philosophy of Science and Historical Enquiry* (Oxford: Clarendon Press, 1987), examines this problem by exploring the relationship between philosophy of science and history of science. In "The Relations between the History and the Philosophy of Science," in *The Essential Tension* (Chicago: University of Chicago Press, 1977), 3–20, Thomas S. Kuhn argues that "active discourse" and not marriage of the two is more useful.

Two of the classic sources in the philosophy of science that still offer much of value are Karl Popper, *The Logic of Scientific Discovery* (New York: Harper and Row, 1959), a difficult but highly rewarding polemic on the nature and philosophy of science, and Popper's companion volume, *Conjectures and Refutations: The Growth of Scientific Knowledge* (New York: Basic Books, 1962).

The classic study of the fundamental metaphysical assumptions on which modern science stands is Edwin Arthur Burtt, *The Metaphysical Foundations of Modern Physical Science*, rev. ed. (New York: Humanities Press, 1951). During the 1950s and '60s, physicist-philosopher Thomas Kuhn became one of the most influential contemporary historians and philosophers of science, writing a number of very significant essays and books, including his highly acclaimed *Structure of Scientific Revolutions* (Chicago: University of Chicago Press, 1962). Drawing heavily on case examples from the sciences, Kuhn provides a theory that accounts for change (although not progress) in science. Kuhn's views have been widely seen as a brilliant contribution to understanding the growth and nature of science, as well as equally criticized for their irrationalist tendencies. See, respectively, Gary Gutting, ed., *Paradigms and Revolutions: Applications and Appraisals of Thomas Kuhn's Philosophy of Science* (Notre Dame: University of Notre Dame Press, 1980), and D. C. Stove, *Popper and After: Four Modern Irrationalists* (New York: Pergamon, 1984).

More recently, Larry Laudan, beginning in his *Progress and Its Problems: Towards a Theory of Scientific Growth* (Berkeley: University of California Press, 1978), has offered perhaps the most compelling explanation for the growth and development of science by arguing that only postpositivist (i.e., historicist) theories of scientific change are capable of providing explanations for scientific progress. Drawing heavily on case studies from the history of science, Laudan continues his discussion in his *Science and Values* (Berkeley: University of California Press, 1984) and in Arthur Donovan, Larry Laudan, and Rachael Laudan, eds., *Scrutinizing Science: Empirical Studies of Scientific Change* (New York: Kluwer Academic, 1988). Although philosophically technical, Laudan is highly valuable and immensely rewarding. In the opinion of the present author, this is the best single discussion of the technical problems involved in understanding science.

The works of Kuhn and Laudan are complemented by other studies arguing for a close connection between the history of science and the philosophy of science that can be consulted for profit. Most recently, two of the more compelling discussions of historicist philosophy of science are Ronald N. Giere, *Explaining Science: A Cognitive Approach* (Chicago: University of Chicago Press, 1988); and David L. Hull, *Science as a Process: An Evolutionary Account of the Social and Conceptual Development of Science* (Chicago: University of Chicago Press, 1988). The following may also be consulted with profit: Stephen Toulmin, *Human Understanding* (Oxford: Clarendon Press, 1972); Gerald Holton, *Thematic Origins of Scientific Thought* (Cambridge, Mass.: Harvard University Press, 1973); Wolfgang Stegmueller, *The Structure and Dynamics of Theories* (Berlin: Springer-Verlag, 1976); Gerald Holton, *The Scientific Imagination: Case Studies* (New York: Cambridge University Press, 1978); Imre Lakatos, *Philosophical Papers*, Vol. I. *The Methodology of Scientific Research Programmes* (New York: Cambridge University Press, 1978); Paul Feyerabend, *Rationalism, Realism and Scientific Method: Philosophical Papers*, Vol. I (New York: Cambridge University Press, 1981); and I. Bernard Cohen, *Revolution in Science* (Cambridge, Mass.: Harvard University Press, 1985).

# Index

# Note on the Author

ERICH ROBERT PAUL, professor of history of science and of computer science at Dickinson College in Pennsylvania, is also the author of *The Milky Way Galaxy and Statistical Cosmology, 1890–1924.*